OTHER BOOKS BY STANLEY BADZINSKI, JR.

Carpentry in Commercial Construction

Carpentry in Residential Construction

House Construction: A Guide to Buying, Building, and Evaluating

Roof Framing

Roof Framing

STANLEY BADZINSKI, JR.

Milwaukee Area Technical College
Milwaukee, Wisconsin

PRENTICE-HALL, INC., Englewood Cliffs, New Jersey

Library of Congress Cataloging in Publication Data

Badzinski, Stanley.
Roof framing.

Includes index.
1. Roofs - - Handbooks, manuals, etc. I. Title.
TH2393.B3 694'.27 76-10810
ISBN 0-13-782466-1

10 9 8 7 6 5 4 3 2 1

Printed in the United States of America

Prentice-Hall International, Inc., *London*
Prentice-Hall of Australia Pty. Limited, *Sydney*
Prentice-Hall of Canada, Ltd., *Toronto*
Prentice-Hall of India Private Limited, *New Delhi*
Prentice-Hall of Japan, Inc., *Tokyo*
Prentice-Hall of Southeast Asia Pte. Ltd., *Singapore*

Contents

Roof Framing

ONE

Rafter Framing Principles

The roof protects the building and its occupants from the effects of weather, but it also is an architectural feature that gives the building a desired appearance. It must be built in accordance with local building codes and the building plans provided by the designer or architect. Carpenters who become proficient in rafter layout are respected by others in the trade and often earn a premium.

KINDS OF ROOFS

The layout of rafters requires the use of the carpenter's steel square, a knowledge of geometry, and the application of trade arithmetic. There are several kinds of roofs that the carpenter may be called upon to lay out and build. Included are the gable roof, the shed roof, the hip roof, the gambrel roof, the intersecting roof, and the mansard roof.

The *gable roof* slopes in two directions and is perhaps the most commonly used (see Fig. 1-1). Gable roofs are usually symmetrical. All rafters in this type of roof are identical and are called *common rafters*. All of the basic rules of roof framing are based on a cross section of a gable roof.

The *shed roof* is a simple design which slopes in one direction (see Fig. 1-2). It is usually used on sheds, residential addi-

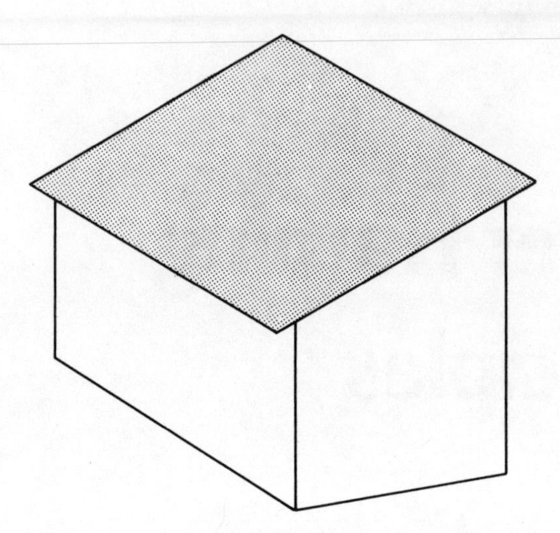

Figure 1-1
Shed roof.

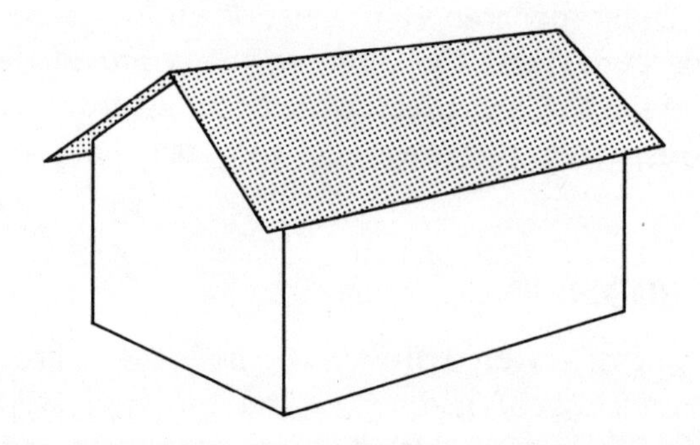

Figure 1-2
Gable roof.

tions, and dormers. All rafters in a given shed roof are identical and are called *common rafters.* However, they differ from common rafters used in the gable roof in that they have a seat cut at each end of the rafter. These seat cuts allow the rafter to rest firmly on top of the walls.

The *hip roof* gets its name from the "hips" at each corner of the building (see Fig. 1-3). This roof slopes in four directions and,

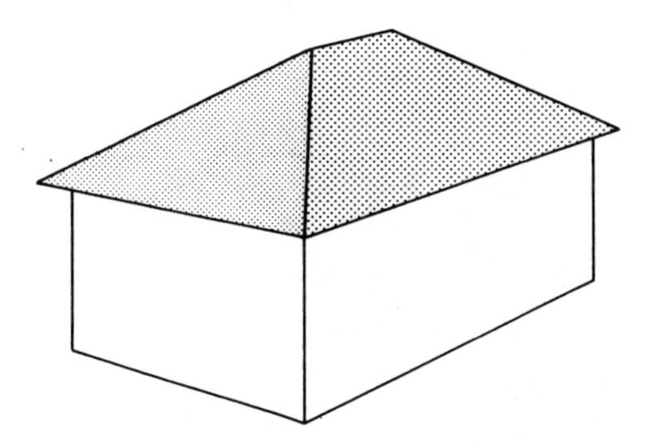

Figure 1-3
Hip roof.

when placed on a rectangular building, contains some common rafters. A hip roof built on a square building is sometimes called a *pyramid roof* (see Fig. 1-4). In the pyramid roof, the hip rafters meet at a common point at the center of the building.

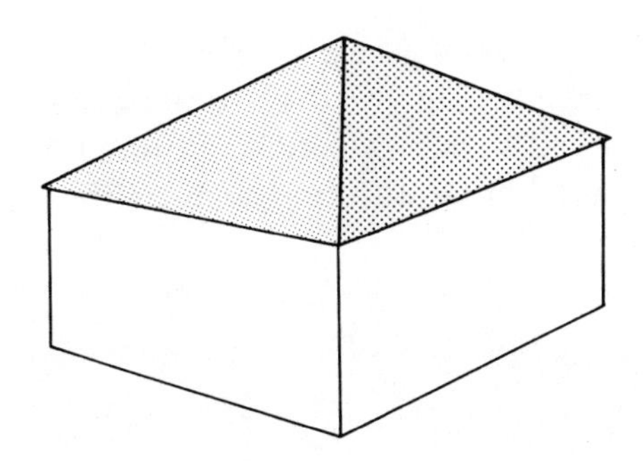

Figure 1-4
Pyramid roof.

When it is necessary to gain the greatest possible use of upper floor space, a *gambrel roof* is sometimes used. This roof has four slopes and is made up of two separate sets of common rafters (see Fig. 1-5). The upper set of rafters is set comparatively flat, and the lower set of rafters is relatively steep. The manner in which the rafters are fastened varies with the design and intended use of the building. The gambrel roof is sometimes called a *barn roof*.

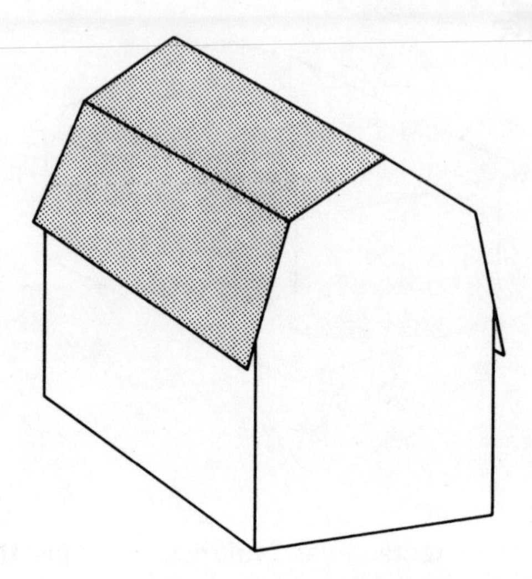

Figure 1-5
Gambrel roof.

The *intersecting roof* is used when the building has a wing or intersecting bay (see Fig. 1-6). The intersection creates a valley on the roof, and valley rafters may be required. Intersecting roofs may be built a number of different ways, either with or without valley rafters.

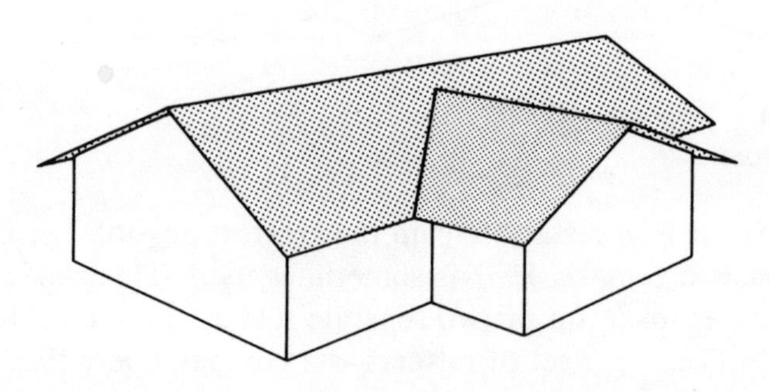

Figure 1-6
Intersecting roof.

The *mansard roof* combines the features of the gambrel and hip roofs and may be built in a number of ways (see Fig. 1-7). The lower portion of the roof is very steep and may be nearly vertical. The upper portion may be a flat deck, or it may be built with a comparatively small slope.

The gambrel roof and the mansard roof incorporate the features of gable and hip roofs. They will not be covered separately in this text because anyone who has mastered the basic gable and hip roofs will have no difficulty in framing either a gambrel or a mansard roof.

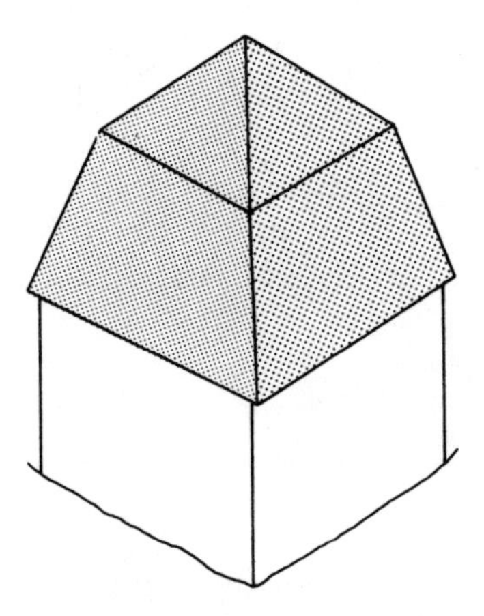

Figure 1-7
Mansard roof.

KINDS OF RAFTERS

Anyone doing roof framing work should be able to identify the various kinds of rafters he encounters. A typical roof will contain either some or all of the following: common rafters, ridge board, hip rafters, hip jack rafters, valley rafters, valley jack rafters, and cripple jack rafters (see Fig. 1-8).

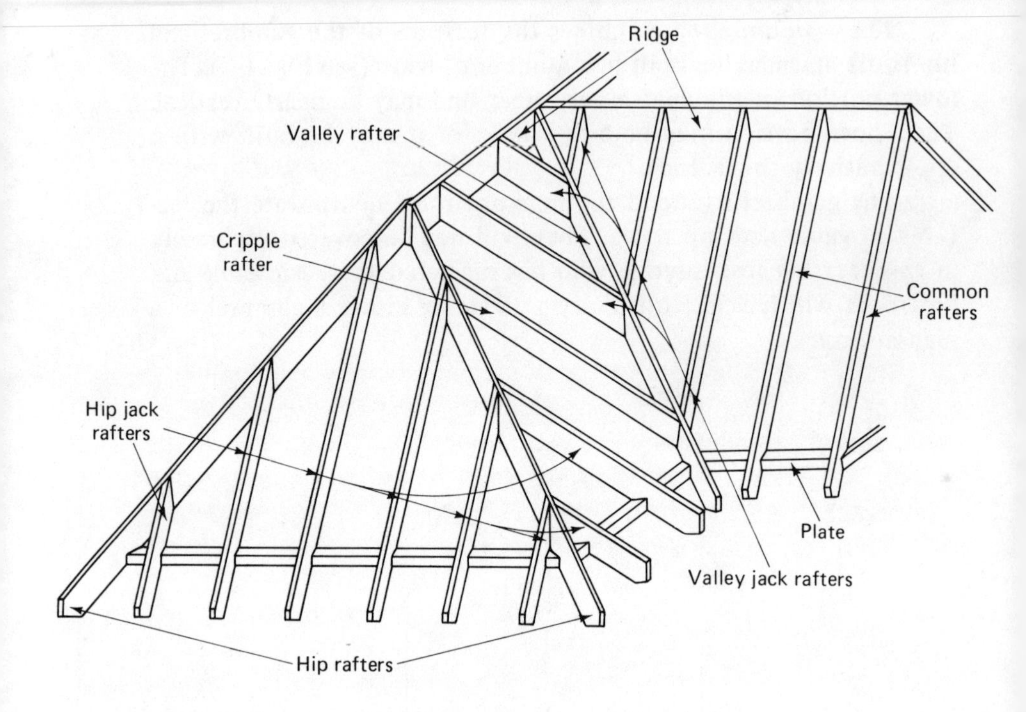

Figure 1-8
Kinds of rafters.

Common rafters run from the top of the wall or plate to the ridge board or peak of the roof. They are the only kind of rafter used in a gable roof.

The *ridge board* is used to hold the common rafters at the proper spacing. It is an aid in erecting common rafters, but in some cases it is not used.

Hip rafters are placed at the outside corners of the building and in plan view run at a 45° angle from the top of the wall to the ridge or peak of the roof.

Hip jack rafters are parts of common rafters that have been cut off at an angle to fit against the hip rafter. They run from the top of the wall to the hip rafter.

Valley rafters are placed at inside corners formed by the intersection of outside walls. They run from the top of the wall to the ridge or peak of the roof.

Valley jack rafters are parts of common rafters that have been cut to fit against the valley rafter. Valley jack rafters run from the valley rafter to the ridge or peak of the roof.

The *cripple jack rafter* is a short part of common rafter cut at an angle on both ends to fit between a hip rafter and a valley rafter. Cripple jacks touch neither the wall nor the ridge.

ROOF FRAMING TERMS

Rafter framing is based on the right triangle, and the basic rules of rafter framing are based on the cross section of a gable roof. To better understand the terminology of roof framing, study the cross section of the roof in Fig. 1-9 in which all parts are represented by lines. Following are definitions of terms used in roof framing:

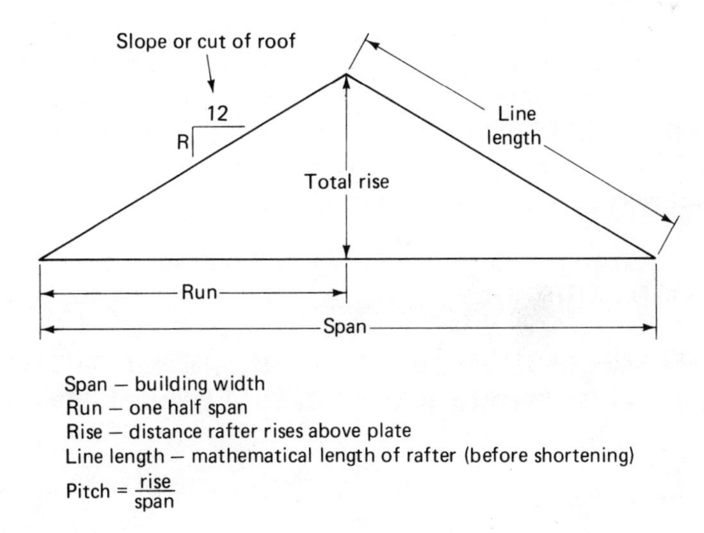

Figure 1-9
Roof framing terms.

Span. The horizontal distance covered by a pair of common rafters. In roof framing, span is equal to the width of the building and is one of the first items the roof framer must know before he can lay out the rafters.

Run. The horizontal distance covered by the common rafter. It is equal to one-half the building span. Common rafter run is determined by dividing the building span by 2.

Total rise. The vertical distance that the rafters rise above the wall. It will vary with the angle of roof incline and the width of the building.

Line length. The mathematical or theoretical length of the rafter. It is the length of the rafter based on a line and does not consider the thickness of rafter stock or framing details.

Pitch. The ratio between rise and span. A roof with a total rise of 5′ and a total span of 20′ would have a pitch of 5/20 or 1/4 pitch. When a fractional pitch is given on the plans, the total rise is determined by multiplying the span by the fractional pitch.

Slope. The slope of the roof is the ratio of rise and run. The slope triangle gives a vertical distance in relation to a horizontal distance. The numbers given by the slope triangle are used on the framing square to lay out the angles on the rafters and are often referred to as the *cut of the roof.*

With an understanding of these basic terms the apprentice is reading to learn more about roof framing. There are a number of different methods in which rafters can be laid out. However, each method makes use of the right triangle in some way. Therefore, a basic knowledge of the right triangle is in order.

THE RIGHT TRIANGLE

The sides of a right triangle are often labeled the base, the altitude, and the hypotenuse (see Fig. 1-10). As applied to a gable

Figure 1-10
Right triangle.

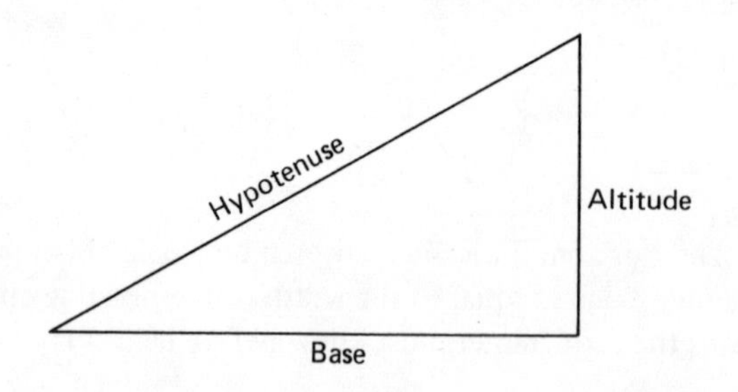

roof, the run of the rafter is the base of the triangle, the rise is the altitude, and the line length is the hypotenuse (see Fig. 1-11).

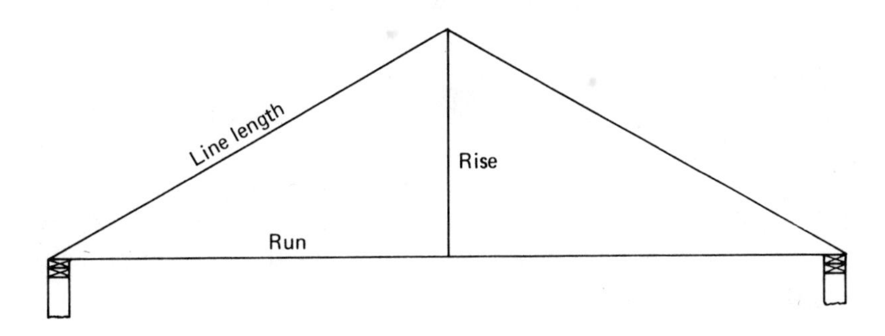

Figure 1-11
Right triangle applied to a gable roof.

The length of the hypotenuse can be easily determined if the length of the other two sides is known. One method is to set the length of the run or base on the body of the framing square, set the length of the rise or altitude on the tongue of the square, and scale the length of the hypotenuse (see Fig. 1-12). In this example,

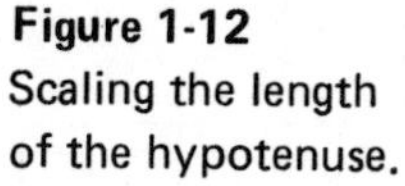

Figure 1-12
Scaling the length of the hypotenuse.

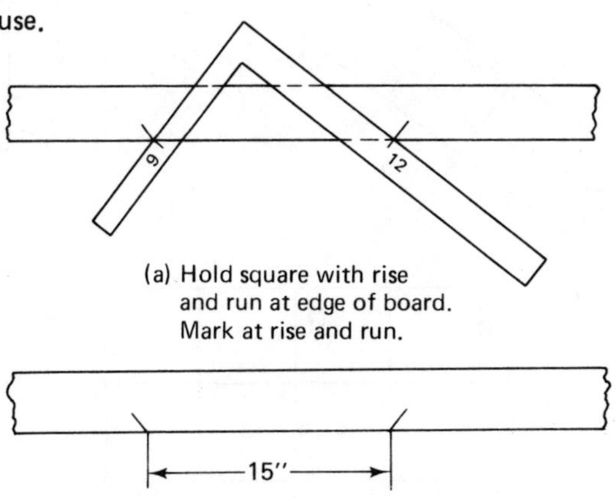

the run was 12″, and the rise was 9″. The diagonal distance from 12″ to 9″ scales 15″. If 12″ and 9″ are used to represent 12′ and 9′ respectively, then the 15″ scaled would represent a length of 15′.

The length of the hypotenuse can be determined mathematically by applying the Pythagorean theorem which states that the sum of the squares of the legs of a right triangle equals the square of the hypotenuse. Applying this theorem in a right triangle that has legs of 3″ and 4″, we find that 3 squared (9) plus 4 squared (16) equals 25. Therefore, the square of the hypotenuse is 25 (see Fig. 1-13), and since 5 times 5 is 25, the length of the hypotenuse is 5, the square root of 25.

Figure 1-13
Squares or sides of right triangle.

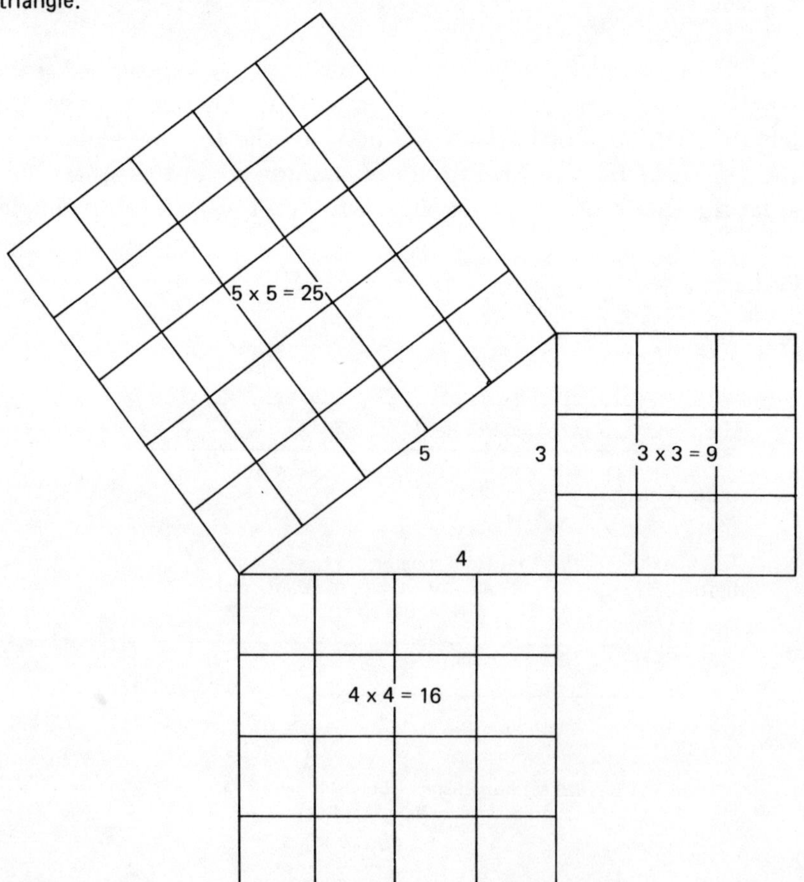

The square root of a given number (25) is a number (5) which when multiplied by itself results in a product of the given number (25). Determining the square root of a number is not always as easy as it was in the previous example. The square root of 225 is 15. It can be determined by trial and error (or by using an electronic calculator), but it can also be obtained arithmetically. The system of determining square root arithmetically seems long and complicated at first, but it is actually easy to work.

EXAMPLE:

Find the square root of 225.

1. Mark off two places from the decimal point in each direction.

   $\sqrt{2'25.00'00}$

2. Take the largest square out of the first group of numbers. The largest square in 2 that is less than 2 is 1. One squared is 1. This 1 is placed below 2 and is subtracted from it.

   $$\begin{array}{l} \phantom{\sqrt{}}1 \\ \sqrt{2'25.00'00} \\ \phantom{\sqrt{}}\underline{1} \\ \phantom{\sqrt{}}1 \end{array}$$

3. Next, the next two digits (25) are brought down and placed along side the remainder (1).

   $$\begin{array}{l} \phantom{\sqrt{}}1 \\ \sqrt{2'25.00'00} \\ \phantom{\sqrt{}}1 \\ \phantom{\sqrt{}}\boxed{1\ 25} \end{array}$$

4. The next step is to determine a trial divisor. This is done by multiplying the number in the quotient (1) by 20 and placing it along side the last row of digits (125).

   $$\begin{array}{ll} & \phantom{\sqrt{}}1 \\ & \sqrt{2'25.00'00} \\ & \phantom{\sqrt{}}1 \\ 20 & \phantom{\sqrt{}}\boxed{1\ 25} \end{array}$$

5. The trial divisor will go into 125 six times, but when we add the 6 to 20, we get 26, and 6 times 26 is greater than 125. Therefore, 5 is placed in the quotient and it is also added to the trial divisor.

   $$\begin{array}{ll} & \phantom{\sqrt{}}1\ \ 5. \\ & \sqrt{2'25.00'00} \\ & \phantom{\sqrt{}}1 \\ 20 & \phantom{\sqrt{}}\boxed{1\ 25} \\ \underline{\ 5} & \\ 25 & \end{array}$$

6. Now the actual divisor (25) is multiplied by the 5 in the quotient. The product (125) is written below the 125 in the bracket. Since there is no remainder, the problem is complete, and we have 15 as the square root of 225.

```
              1  5.
           √2'25.00'00
              1
    20     | 1 25
     5     | 1 25
    25
```

To continue the example, let's find the square root of 225.879. The first six steps are the same as in the preceding example except that there are significant numbers to the right of the decimal point.

7. After subtracting 125 from 125 we bring down the next group of two digits (87), multiply the number in the quotient (15) by 20, and write a new trial divisor (300) to the left of 87 in the problem.

```
              1  5.
           √2'25.87'90
              1
    20     | 1 25
     5     | 1 25
    25      300 | 87
```

8. Since 300 will not go into 87, we write 0 in the quotient and add it to the trial divisor. Multiplying the divisor by the last number in the quotient (0), we write 00 below the 87, subtract, and bring down the next two digits (90).

```
              1  5. 0
           √2'25.87'90
              1
    20     | 1 25
     5     | 1 25
    25      300 | 87
              0 | 00
            300 | 87 90
```

9. Next, we create a new trial divisor by multiplying the quotient (150) by 20. Since new trail divisor, 3,000, will go into 8,790 two times, we write 2 in the quotient, add it to the trial divisor, and follow through with steps 6 and 7.

```
              1  5. 0  2
            √2'25.87'90
              1
   20        1 25
    5        1 25
   25   300      87
          0
        300      00
       3000      87 90
          2
       3002      60 04
                 27 86
```

10. Because of the large remainder we may decide to continue on by adding on additional groups of double zeros, and continue as we did in step 9.

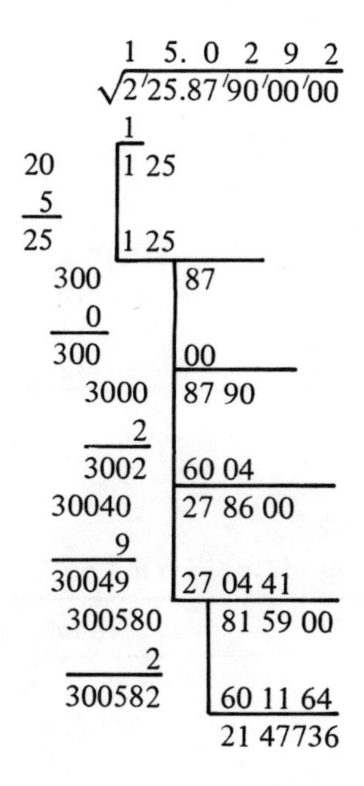

Although there is no limit to the number of decimal places, it is not necessary to determine the square root of a number beyond the nearest 100th. Many applications in roof framing require determining the length of the hypotenuse of a triangle. The length can be determined quickly and reasonably accurately by scaling. Where greater accuracy is required, the Pythagorean theorem should be used. Both methods will be discussed in detail, as required, in the following chapters.

Table 1-1

Inches to Feet in Decimals

Inches	*Feet*
1	.08
2	.17
3	.25
4	.33
5	.42
6	.50
7	.58
8	.67
9	.75
10	.83
11	.92
12	1.00

Table 1-2

Fractions of One Inch to Feet in Decimals

Inches	*Feet*
1/16	.005
1/8	.01
1/4	.02
3/8	.03
1/2	.04
5/8	.05
3/4	.06
7/8	.07
1	.08

CONVERSION FACTORS

Everyone knows that there are 12″ in 1′ but many are confused when they have to convert inches to a decimal part of a foot. Therefore, the following explanation of converting inches to a decimal part of a foot is included here.

One inch equals one-twelfth of a foot. Expressed mathematically, 1″ = 1/12′. If 1 is divided by 12, the result is .0833. With 1″ being .0833′, 2″ is .1666′. Because it is common practice to work to the nearest 1/100, .1666 is rounded off to .17′. In a similar manner, .0833 is rounded off to .08′. Table 1-1 gives the decimal foot equivalent for whole inches. The table was prepared by adding and subtracting .08 to each quarter. In this way it is possible to avoid the accumulation of error that would occur if only .08 were added to each proceeding inch.

When fractions of 1″ must be converted to a decimal part of a foot, it can be assumed (since 1″ equals .08′) that 1/8″ equals .01′. It is seldom necessary to work in increments smaller than 1/8″, but if necessary, 1/16″ can be set at .005 (see Table 1-2).

REVIEW QUESTIONS

1. List two functions of a roof.
2. Name the three areas of knowledge that are used to lay out roofs.
3. List five common kinds of roofs.
4. What kind of rafters are used in gable roofs?
5. What kinds of rafters are used in hip roofs?
6. What kinds of rafters are used in intersecting roofs?
7. Define the following:

   Span
   Run
   Total rise
   Line length
   Pitch
   Slope

8. Find the square root of the following:
   225
   196
   180
   160
   153

TWO

Gable Roofs and Shed Roofs

The gable roof and the shed roof are among the easiest to build because all the rafters are the same. Of the two kinds the gable roof is more commonly used, and the basic rules of roof framing relate to a cross section of the gable roof.

THE GABLE ROOF

The gable roof is made of pairs of common rafters that meet at the ridge (see Fig. 2-1). The rafter is laid out in two parts: the body and the tail. The body lies over the building and is the most important part of the rafter. The length of the body is called *line length.*

The length of the tail that projects outside the wall is called *overhang.* The horizontal distance covered by the tail is called *projection* or *projection of run* (see Fig. 2-1).

The line length is based on the diagonal of the rise and the run which form the theoretical triangle (shaded area Fig. 2-1). The theoretical line falls somewhere between the rafter edges, but it is not necessary to know the exact location of this line because the line length is marked on the top edge of the rafter stock.

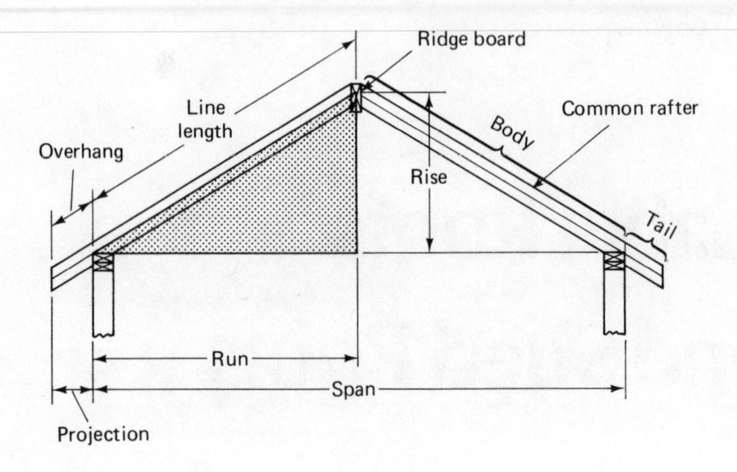

Figure 2-1
Common rafter terminology.

This can be done without changing the length of the rafter because the edges of the rafter stock are parallel to the theoretical line length, and the ends of the rafter are marked with plumb lines. *Plumb lines* are lines that are plumb or vertical when the rafter is put into position, and since all vertical lines are parallel, the length of the rafter is the same whether measured at the top or on a line drawn parallel to the top edge (see Fig. 2-2).

Figure 2-2
Common rafter length marking.

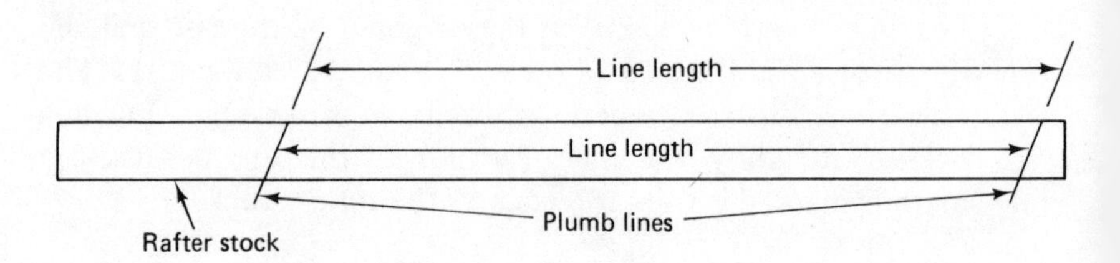

Common Rafter Length

Common rafter length may be determined in a number of ways. If only the width of the building and the total rise of the roof are known, the rafter length may be determined either mathematically or by scaling. In Fig. 2-3 the span is given as 26′, and the total rise is given as 8′–8″. The first step is to determine the

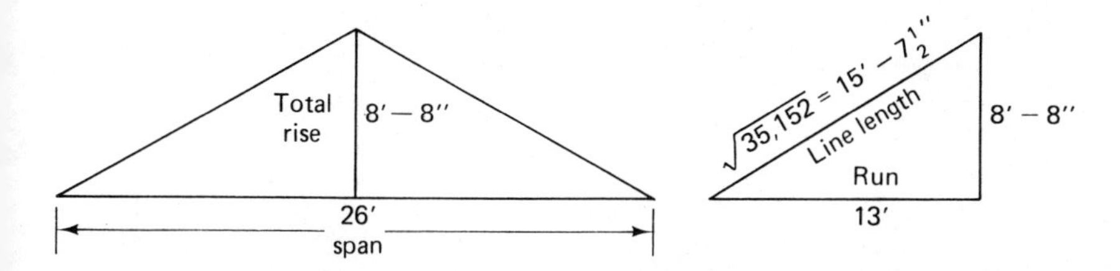

Figure 2-3
Common rafter problem.

run of the rafter. Then, by using a sketch of the rafter problem, the total rise and total run are marked on legs representing the rise and run. In some cases, it is advantageous to change the unit dimensions from feet to inches. If this is done, the same units must be used for both rise and run.

EXAMPLE:

Total rise	8′–8″	= 104″
Total run	13′	= 156″

Applying the Pythagorean theorem, we get

$$(\text{Rafter length})^2 = (156)^2 + (104)^2$$

$$(\text{Rafter length})^2 = 24{,}336 + 10{,}816$$

$$\text{Rafter length} = \sqrt{35{,}152}$$

NOTE: For a discussion on obtaining the square root of a number, see Chapter 1.

$$\text{Rafter length} = 187.49''$$

At this point the rafter length in inches and a decimal fraction must be changed to feet and inches. To do this, the inches are divided by 12, but they are not divided beyond the decimal point.

$$\begin{array}{r} 15' \\ 12\overline{)187.49} \\ \underline{12} \\ 67 \\ \underline{60} \\ 7.49'' \end{array}$$

The answer at this point is $15'-7.49''$. Since the length of the rafter is desired to the nearest 1/16", .49" must be changed to a common fraction. To change a decimal to the nearest 1/16, the decimal is multiplied by 16/16, and the result in the numerator is rounded off to the nearest whole number.

$$.49'' \times \frac{16}{16} = \frac{7.84}{16} = \frac{8}{16} = \frac{1''}{2}$$

Therefore, the mathematical length of the common rafter becomes $15'-7\frac{1}{2}''$ and can easily be marked on the common rafter stock as illustrated in Fig. 2-8.

Scaling Common Rafter Length. On the back of the rafter framing square the inches on the outer edge are divided into twelfths. By setting 1" equal to 1' the total rise may be measured on the tongue of the square and the total run on the body of the square. To determine the diagonal distance between the total rise and total run, the square is held along the edge of a straight board and marks are placed on the board as shown in Fig. 2-4.

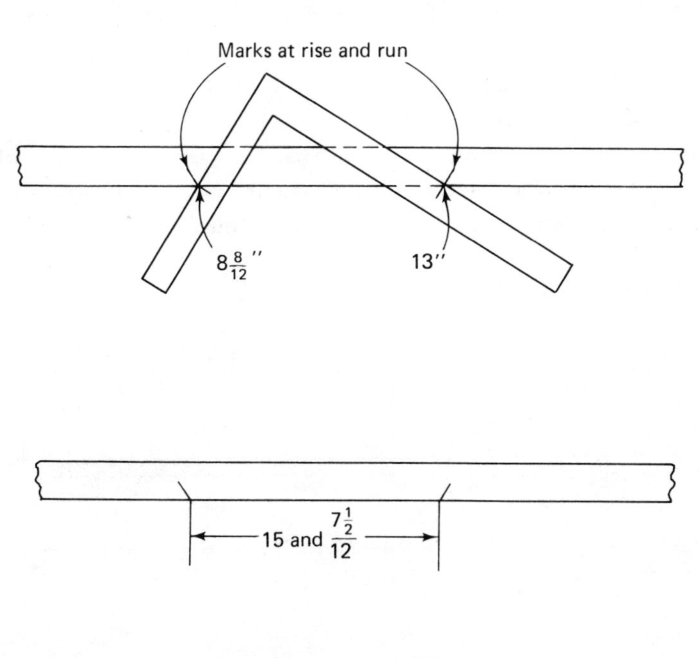

Figure 2-4
Scaling common rafter length.

Care must be exercised in holding the square so that the total rise and total run dimensions are exactly in line with the edge of the board. Any deviation from the exact dimensions will cause an error in the rafter length. Care must also be exercised in making the length marks. A heavy pencil or carelessly held pencil can add inches to the rafter length. For this reason it is desirable to mark the length with a pocket knife.

After the marks have been made, the distance between the marks is measured by using the 1/12 scale on the framing square.

In Fig. 2-4 the distance between the marks accurately measured is $15\frac{7.5}{12}''$. When converted to feet, it represents a rafter length of $15'-7\frac{1}{2}''$.

The scale method of finding rafter length is theoretically correct. It can be used to quickly determine the approximate length of a rafter. With careful marking and measuring it will give the accuracy necessary for preparing rafter patterns.

If the 1/12 scale is used, the rafter length may also be determined by the step method. By allowing 1" for every foot of rise and run the numbers held on the square actually represent 1/12 the total rise and total run. Therefore, marking 12 consecutive distances between rise and run along the edge of the rafter stock yields the rafter line length (see Fig. 2-5).

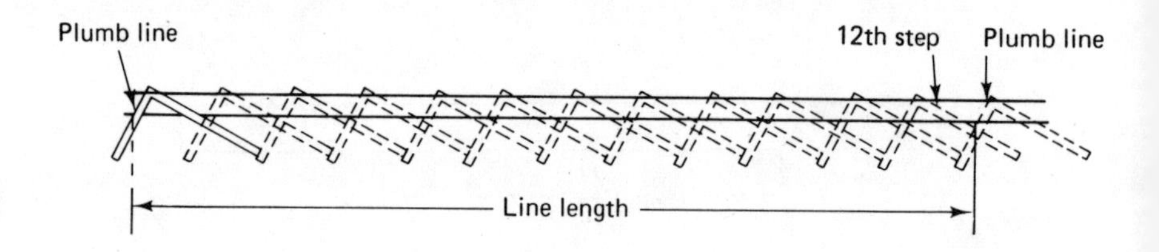

Figure 2-5
Scale step-off common rafter length.

Stair gages are useful when stepping the rafter length. The gages are positioned on the square so that when the square is held on the rafter stock the rise and run are exactly at the edge of the stock. Starting at one end of the rafter stock a plumb mark is made along the rise side of the square and a small mark is made on the run side. The square is then slipped along the edge of the rafter stock until the rise side of the square is exactly in line with the previous mark. A mark is again made on the run side of the square. The square is slipped along the edge of the rafter stock until 12 steps have been made and the rafter length determined. A plumb line is drawn at the twelfth step to indicate the line length (see Fig. 2-5).

Unit Method Rafter Length. The unit method of rafter layout is an established simple method for determining rafter length. The basic unit of run is 12″ or 1′. The unit run for any common rafter or jack rafter is always 1′. When working with total rise and total run, the total run was equal to one-half the span. The unit run is equal to one-half the unit span. Therefore, the *unit span* is 2 × 12″ or 24″ (see Fig. 2-6).

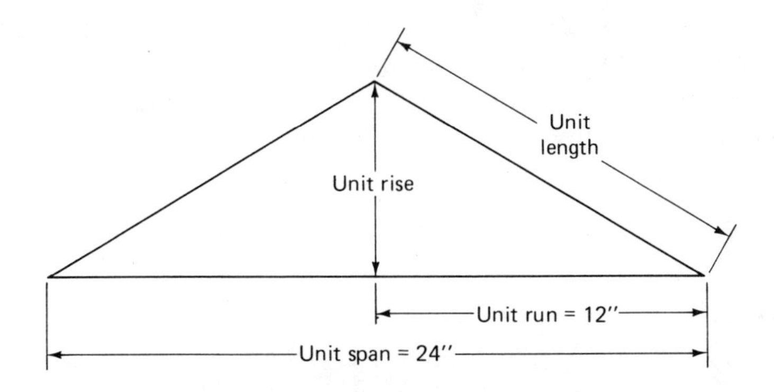

Unit span – always = 24″
Unit run – always = 12″
Unit rise – changes with slope or pitch
Unit length – increases with rise

Figure 2-6
Roof framing units.

The unit rise or rise per foot of run is selected to meet the requirements of the roof and the overall design of the building. A small unit rise is used for comparatively flat roofs. Larger unit rises are used for steeper roofs.

The unit length is the line length of the rafter for 1′ of run and 1 unit of rise. The unit length for unit rises from 2″ to 18″ is given on the rafter table of the steel square (see Fig. 2-7). These unit lengths were determined by applying the Pythagorean theorem. The inch marks on the outer edge of the square represent the unit rise, and the line length of the rafter is given to the nearest .01″ on the first line of the rafter table. This table makes it possible for the carpenter to determine rafter lengths accurately without having to extract the square root of a number.

In preparing to lay out rafters, the unit rise is usually determined by consulting the wall cross section in the building plans for the roof slope triangle, but sometimes either a fractional pitch or a total rise is given.

When the slope triangle is used, the horizontal leg of the slope triangle is always 12″ which represents one unit of run. The vertical leg represents the unit rise and varies.

When only a fractional pitch is listed, the unit rise may be determined by applying the pitch formula:

$$\text{Pitch} = \frac{\text{rise}}{\text{span}}$$

EXAMPLE:

$$P = \frac{1}{3}$$

$$\text{Pitch} = \frac{\text{unit rise}}{\text{unit span}}$$

$$\frac{1}{3} = \frac{\text{unit rise}}{24''}$$

$$\text{Unit rise} = 8''$$

Once the pitch formula is understood, there is no need to work it in the manner shown in the previous example. All that is necessary is to multiply the fractional pitch by the unit of span.

EXAMPLE:

$$P = \frac{1}{3} \qquad \text{Unit span} = 24''$$

$$\text{Unit rise} = P \times 24''$$

$$\text{Unit rise} = \frac{1}{3} \times 24 = \frac{24}{3} = 8''$$

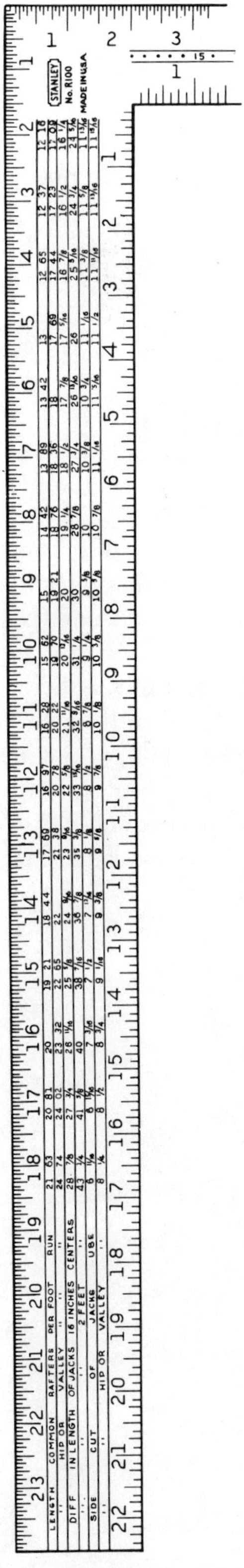

	18	17	16	15	14	13	12	11	10	9	8	7	6	5	4	3	2
LENGTH COMMON RAFTERS PER FOOT RUN	21 63	20 81	20	19 21	18 44	17 69	16 97	16 28	15 62	15	14 42	13 89	13 42	13	12 65	12 37	12 16
" HIP OR VALLEY " " "	24 74	24 02	23 32	22 65	22	21 38	20 78	20 22	19 70	19 21	18 76	18 36	18	17 69	17 44	17 23	17 09
DIFF IN LENGTH OF JACKS 16 INCHES CENTERS	28 7/8	27 3/4	26 11/16	25 5/8	24 9/16	23 9/16	22 5/8	21 11/16	20 13/16	20	19 1/4	18 1/2	17 7/8	17 5/16	16 7/8	16 1/2	16 1/4
" " " 2 FEET "	43 1/4	41 5/8	40	38 7/16	36 7/8	35 3/8	33 15/16	32 9/16	31 1/4	30	28 7/8	27 3/4	26 13/16	26	25 5/16	24 3/4	24 5/16
SIDE CUT OF JACKS USE	6 11/16	6 15/16	7 3/16	7 1/2	7 13/16	8 1/8	8 1/2	8 7/8	9 1/4	9 5/8	10	10 3/8	10 3/4	11 1/16	11 3/8	11 5/8	11 13/16
" " HIP OR VALLEY "	8 1/4	8 1/2	8 3/4	9 1/16	9 3/8	9 5/8	9 7/8	10 1/8	10 3/8	10 5/8	10 7/8	11 1/16	11 5/16	11 1/2	11 11/16	11 13/16	11 15/16

STANLEY
No. R100
MADE IN U.S.A.

Figure 2-7
Rafter tables for roof framing.

When the total rise is given, the unit rise can be determined by dividing the total rise in inches by the number of units of run.

EXAMPLE:

$$\text{Total rise} = 5'\text{–}5''$$

$$\text{Run} = 13'$$

$$5'\text{–}5'' = 65''$$

$$65'' \div 13 \text{ units} = 5'' \text{ unit rise}$$

To calculate the line length of the common rafter, the unit length on the rafter framing table is multiplied by the number of units of run. (Because a unit of run is one foot, there are the same number of units of run as there are feet of run. In multiplying run by unit length, the number of units is being multiplied by unit length. Feet are not being multiplied by inches.) The initial answer gives the rafter length in inches. The total length in inches must be divided by 12, stopping at the decimal point, to get an answer in feet, inches, and fraction of an inch. The decimal fraction is changed to the nearest 1/16″ in the same manner as described on page 20.

EXAMPLE:

$$\text{Span} = 26' \qquad \text{Unit rise} = 4$$

$$\text{Run} = 13' \qquad \text{Unit length} = 12.65''$$

$$\text{Rafter length} = 13 \times 12.65'' = 164.45''$$

$$\text{Rafter length} = 13'\text{–}8.45'' = 13'\text{–}8\tfrac{7}{16}''$$

After the rafter length has been determined, it is marked on the top edge of the rafter stock. The usual procedure is to square a line across the top edge of the rafter stock near one end. Then the line length is measured out and a line is squared across the edge of the rafter stock at the line length mark (see Fig. 2-8).

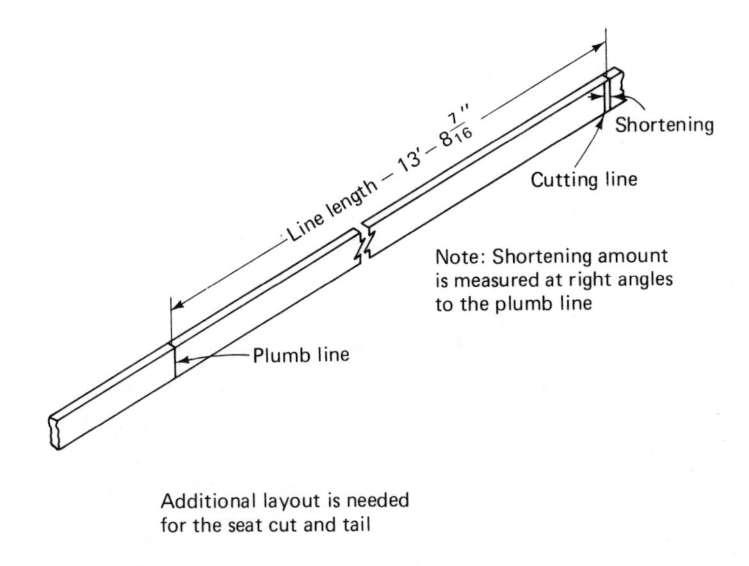

Figure 2-8
Common rafter layout.

A plumb line is drawn on the side of the rafter stock at the length marks. Plumb lines are obtained by holding rise on one side of the square, run on the other side, and marking on the rise side of the square.

To allow for the thickness of the ridge board, the rafter is shortened. The amount of shortening is equal to one-half the thickness of the ridge board. It is deducted by measuring at right angles to the plumb line. A new plumb line is drawn at the shortened distance. This new plumb line is the cutting line.

The layout for the rafter tail and seat cut is made after the line length layout is completed. This layout must be made in conjunction with the cornice detail and is discuused on pages – .

Some carpenters prefer to lay out the common rafter by using the step-off method. In the step-off method, the framing square is held with the unit rise along the edge of the rafter on the tongue of the square and the unit run on the body. A mark is made on each side of the square, and the square is stepped off a number of times equal to the run of the rafter. If the rafter has a fraction of a foot of run, this fraction is stepped off in the usual manner, but instead of placing a mark at the edge of the rafter for a full unit of run, a mark is placed at the inch mark equal to the fraction of the run (see Fig. 2-9).

Although the step-off method of laying out a rafter is easy and theoretically correct, it is not as accurate as the unit method because a small amount can be gained or lost on each step. A gain of only 1/16″ on each of the 12 steps results in a rafter 3/4″ too long. To avoid gaining length on each step, a sharp pencil or knife should be used to mark each step, and the square should be aligned carefully for each step. Stair gages clamped to the square as described on pages 21–22 make it easier to lay out the rafter with the step method.

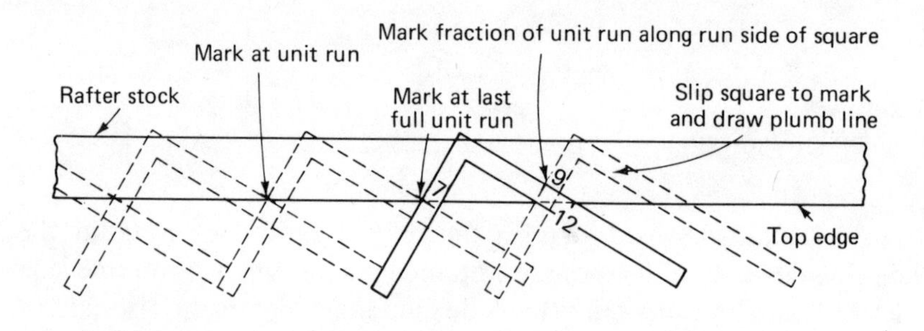

Figure 2-9
Common rafter layout – unit step-off method.

Layout of Common Rafter Tail. The tail of the rafter is laid out in accordance with the cornice detail. Because this detail is drawn to a small scale, it does not give the carpenter all the information he needs. To get the necessary information, the carpenter must make a full-sized drawing of the cornice showing all its parts and the shape of the rafter tail. This drawing is usually made on a sheet of plywood or on the subfloor, but any flat surface can be used. The length of the common rafter tail can be determined by using the unit-length method or the step-off method. The cornice detail in Fig. 2-10 shows a 24″ projection. The length of the

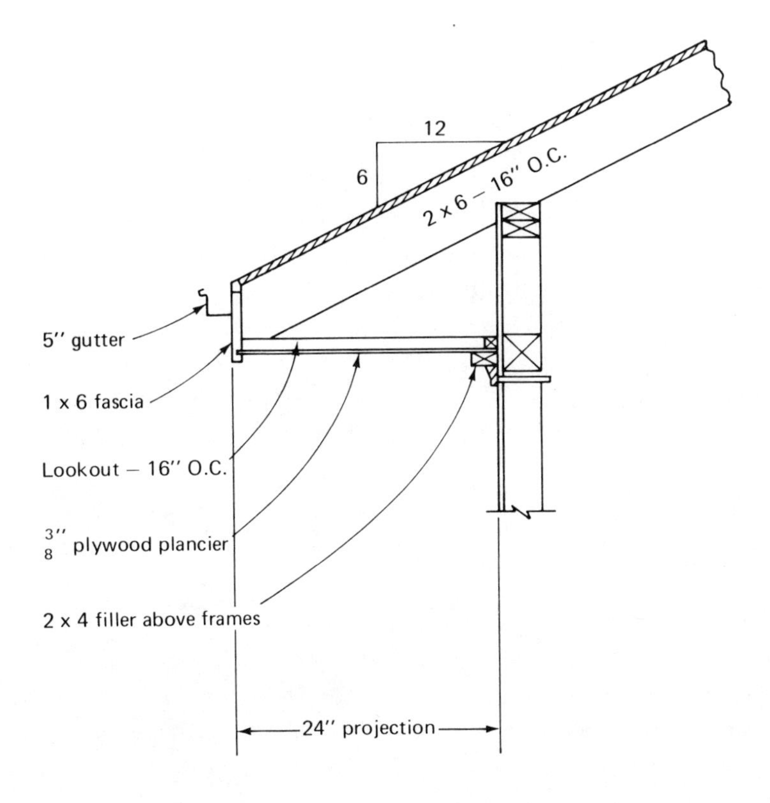

Figure 2-10
Cornice detail.

tail can be found either by multiplying the unit length by 2 and transferring it to the rafter stock or by stepping off the unit rise and unit run two times.

After the length of the tail is determined, the height of the seat cut and the height of the level cut at the end of the tail must be determined. This information is found by measuring the height of the seat cut and tail cut on the full-sized cornice drawing and transferring these dimensions to the common rafter pattern. Figure 2-11 shows the tail of common rafter laid out to meet the requirements of the cornice in Fig. 2-10. A similar layout would be made for any size rafter tail by simply transferring dimensions from a full-sized drawing of the cornice detail.

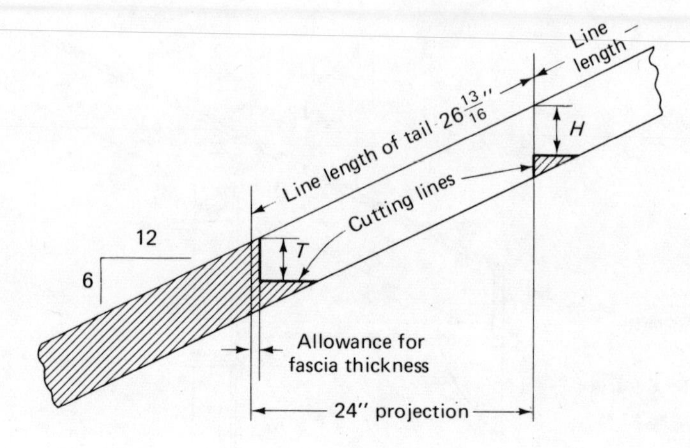

"*H*" is height of seat cut measured along the plumb line from top of rafter to top of wall on full size cornice layout and transferred to rafter pattern

"*T*" is height of tail plumb cut transferred from the full size cornice detail

Figure 2-11
Common rafter — tail layout.

It should be noted that in Fig. 2-10 the level cut in the common rafter seat has full bearing on the top plate. If possible, this level cut should not be longer than the thickness of the wall to which it is framed.

Review of Common Rafter Layout

Before a rafter can be laid out, the span, the run, and the angle of incline must be determined. If the mechanic has information, he can begin to make the rafter layout.

A logical procedure to follow is listed below:

1. Check plans to determine pitch, unit rise, or total rise.
2. Check span by measuring width of building.

3. Divide span by 2 to find run of rafter.
4. Find rafter length.
5. Mark rafter length on rafter stock.
6. Shorten for ridge board.
7. Lay out rafter tail and seat cut.
8. Check rafter length.
9. Check all other layouts.
10. Cut rafter pattern.

SHED ROOF

The shed roof that rests on two walls uses common rafters that have a seat cut at each end. The run of the rafter is the distance from the outside of the lower wall to the inside of the high wall (see Fig. 2-12). The length of the rafter is calculated and laid out in the same manner as for common rafters for a gable roof.

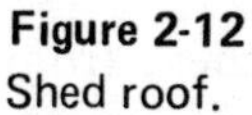

Figure 2-12
Shed roof.

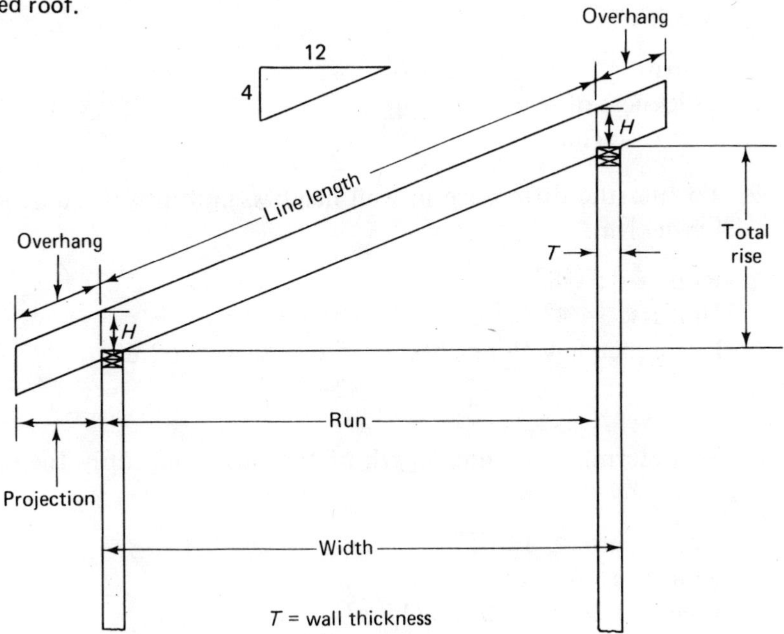

The length of the overhang at each end of the rafter is laid out in conjunction with the cornice detail. Seat cut height must be the same at each end except when it is necessary to adjust for a variation in wall height. If the difference between supporting wall heights is not given on the plan, it may be calculated by multiplying the unit rise by the run of the rafter.

Shed Roof Layout

In the following example used to illustrate rafter layout for a shed roof, the building is 16′ wide, the cut of the roof is 4 and 12, and the projection on each end is 12″. The walls are made of 2 X 4's covered with 1/2″ thick sheathing material (see Fig. 2-12).

Before the rafters are laid out it is necessary to check the width of the building to determine the run and to check the difference in wall heights.

EXAMPLE:

1. To find the run of the rafter, subtract wall thickness from width of building.

Width	16′–0″
Thickness of wall	4″
Run of rafter	15′–8″

2. To find the difference in wall heights, multiply the unit rise by the run.

   Run = 15′–8″ = 15.67′
   Unit rise = 4″
   Difference in wall heights = 15.67 X 4 = 62.68″
   = $62\frac{11}{16}$″

3. To determine the line length of the rafter, multiply the unit length by the run.

   Unit length = 12.65″
   Run = 15.67 units
   Line length = 12.65 X 15.67
   = 198.23 = 16′–$6\frac{1}{4}$″

4. To determine the length of the rafter tails, multiply the unit length by the projection of run.

   Unit length = 12.65″
   Projection = 1′–0″
   Length of tail = 12.65 × 1 = 12.65″ = $12\frac{5}{8}$″

5. To lay out the rafter, mark the length of the upper tail on the top edge of the rafter stock first. Then mark the line length and finally the length of the lower tail on the rafter [see Fig. 2-13(a)]. Draw plumb lines on the side of the rafter at the length marks.

Figure 2-13
Shed roof rafter layout.

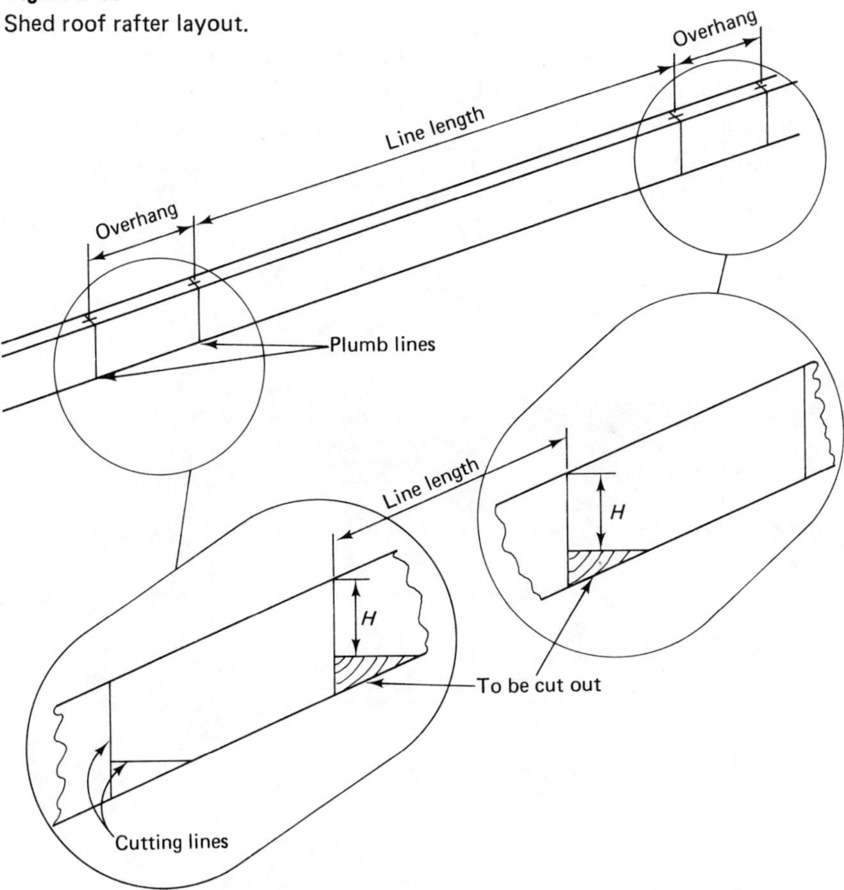

6. To lay out the seat cut and rafter tail, check the cornice detail to determine the height of the seat cut and shape of the tails. If necessary, a full-size drawing may be made to determine the exact dimensions as described on pages 00–00. Transfer the dimensions to the side of the rafter. The layout is now completed [see Fig. 2-13(b)].

Shed Dormers

Roofs with shed dormers are comparatively easy to lay out and build. The cross section of the roof in Fig. 2-14 is used to illustrate the procedure.

The layout of this roof begins with the layout of the dormer wall heights. In this problem the main roof has a total rise of 10′–6″. This total rise was determined by multiplying the unit rise, 9″, by the number of units of run, which is 14 in this example.

The total rise of the dormer roof is found in a similar manner. The unit rise of 3″ is multiplied by 14 units to get a total rise of 42″. The difference in the height of the walls is determined by

Figure 2-14
Shed dormer.

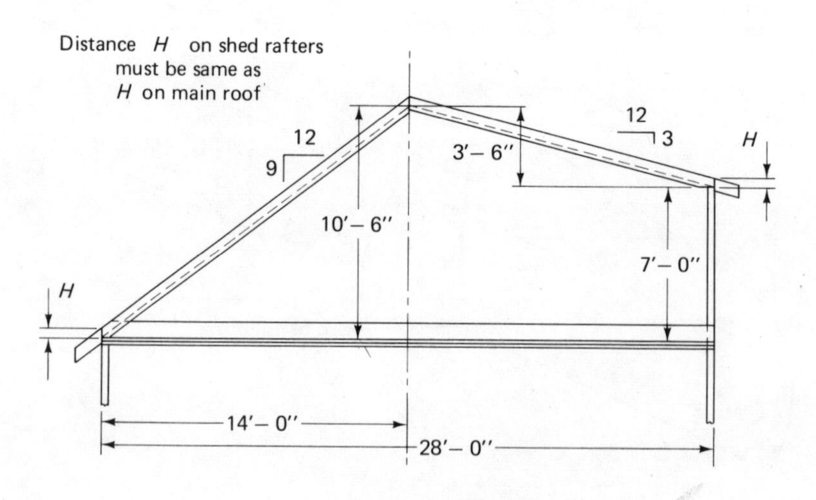

subtracting the total rise of the dormer roof from the total rise of the main roof. In this case, it is 84″ from the top of the wall on which the main rafters rest to the top of the dormer wall.

The line length of the dormer rafter is found by applying the rules for the layout of a common rafter, and the rafter is shortened in the usual manner. When the seat cut is laid out, the height of the seat cut (H in Fig. 2-14) measured along the plumb line from the top of the rafter to the level line must be the same as for the height of the seat cut for the common rafters in the main roof.

REVIEW QUESTIONS

1. Name the two parts of a common rafter.
2. What is overhang? Projection?
3. What is theoretical length?
4. How is theoretical length marked on the common rafter?
5. What are plumb lines?
6. Give three methods of determining common rafter length.
7. How may common rafter length be scaled?
8. Why are stair gages useful in stepping the common rafter length?
9. What is the unit of span?
10. What is the unit of run?
11. What is a unit rise?
12. Calculate the unit rise for the following pitches:

    1/4 pitch — 1/3 pitch
    1/6 pitch — 3/8 pitch
    5/24 pitch — 1/2 pitch

13. Find the unit length for the given unit rises.

    4″
    5″
    6″
    8″
    9″

14. Is the step-off method an accurate way to lay out rafters?
15. Find the length of common rafters for the following conditions.

Unit Rise	*Run*
4″	13′
5″	12′–8″
6″	13′–6″
8″	13′–4″
10″	14′–6″

Pitch	*Span*
5/12	26′
5/24	27′
1/4	28′
1/3	24′–6″
1/6	25′

16. List a logical procedure to follow for laying out common rafters.
17. How is the run of a shed roof determined?
18. How is the seat cut for shed rafters made?
19. How is the length of studs for shed dormers determined?
20. How is H for shed dormer rafters determined?

THREE

Equal Pitch Hip Roofs

Equal pitch roofs have the same slope in all four directions. Hip rafters are placed at the outside corners of the building. These rafters run to the ridge at a 45° angle in plan view. Hip rafters are made from stock 2″ wider than common rafters and jack rafters. This is necessary in order to provide support for the full length of the jack rafter plumb cuts.

ROOF PLANS

Usually, it is not necessary to have a roof plan for a simple hip roof because the location of all the rafters can be easily visualized. However, a roof with a number of offsets requires a roof plan, and a roof plan for a simple roof helps the beginner to understand roof framing.

The plan view shows the location of the rafters by their center lines. The length of the center line in plan view is actually the run of the rafter in a scale drawing. Ordinarily, the plan view shows the outline of the building wall line, the cornice line, ridge boards, hip rafters, and valley rafters if used. Common rafters and jack rafters are usually omitted on the plan drawing, but the carpenter may wish to draw them in to aid in determining the location and length of the various rafters (see Fig. 3-1).

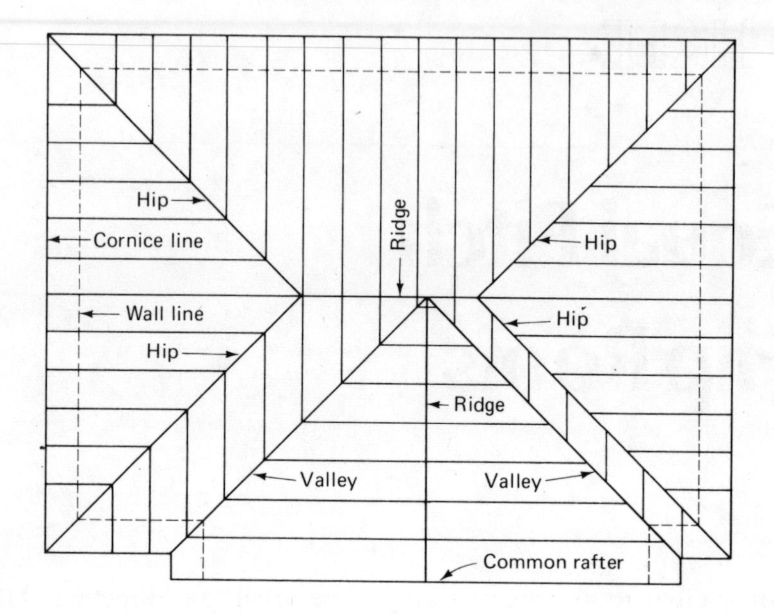

Figure 3-1
Roof plan.

LAYOUT OF RIDGE BOARDS

The ridge board is an aid used in erecting the roof. The locations of the various rafters are marked on the ridge board. Ridge boards for gable roofs are cut to the same length as the building, but ridge boards for hip roofs must be laid out to make allowances for the various lumber thicknesses and the method used to frame the roof.

The roof in Fig. 3-1 has the hip rafters framed directly to the ridge at one end of the building, but the hip rafters at the other end are framed into three common rafters. Because the hip rafters move into the building at a 45° angle in plan view, the theoretical length of the ridge may be determined by subtracting the width of the building from the length of the building.

This theoretical length is measured on the center line of the ridge board and is the distance between the framing points of the hip rafters. The framing points are created by the intersection of center lines and are the points to which all mathematical or theoretical lengths are calculated.

The actual length of the ridge board is longer than the theoretical length (see Fig. 3-2), and the amount added at each end depends on the method used in framing the hip rafters.

At each end where the hip rafters frame into the sides of three common rafters, the ridge board must be increased in length by an amount equal to one-half the common rafter thickness (see Fig. 3-2). When the ridge board is thinner than the common rafter stock, the common rafter framing to the end of the ridge must be shortened more than the other common rafters. The amount of shortening for the end common rafter is one-half the common rafter thickness. All other common rafters are shortened one-half the thickness of the ridge board (see Fig. 3-2).

Figure 3-2
Theoretical and actual ridge length.

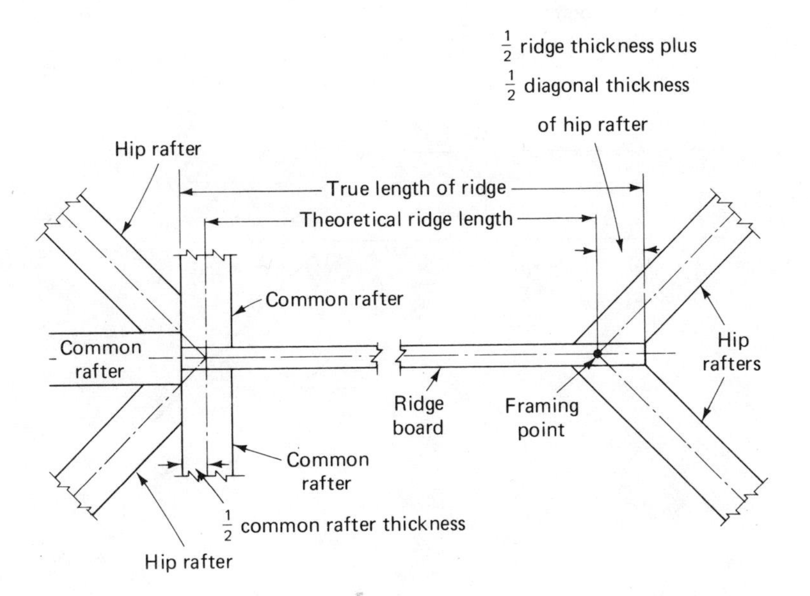

At each end where hip rafters are framed directly to the ridge board, the ridge board must be increased in length by an amount equal to one-half the thickness of the ridge board plus one-half the diagonal thickness of the hip rafter stock (see Fig. 3-3). This total amount of increase in length is necessary because the center line of the hip rafter moves across the top edge of the ridge at a 45° angle. The distance from the framing point to the center line of the hip rafter measured along the edge of the ridge is one-half the ridge thickness. To provide support for the entire bevel cut on the hip rafter, an additional one-half the diagonal thickness of the rafter stock must be added to the theoretical length of the ridge board.

Figure 3-3
Hip rafter framing to ridge board.

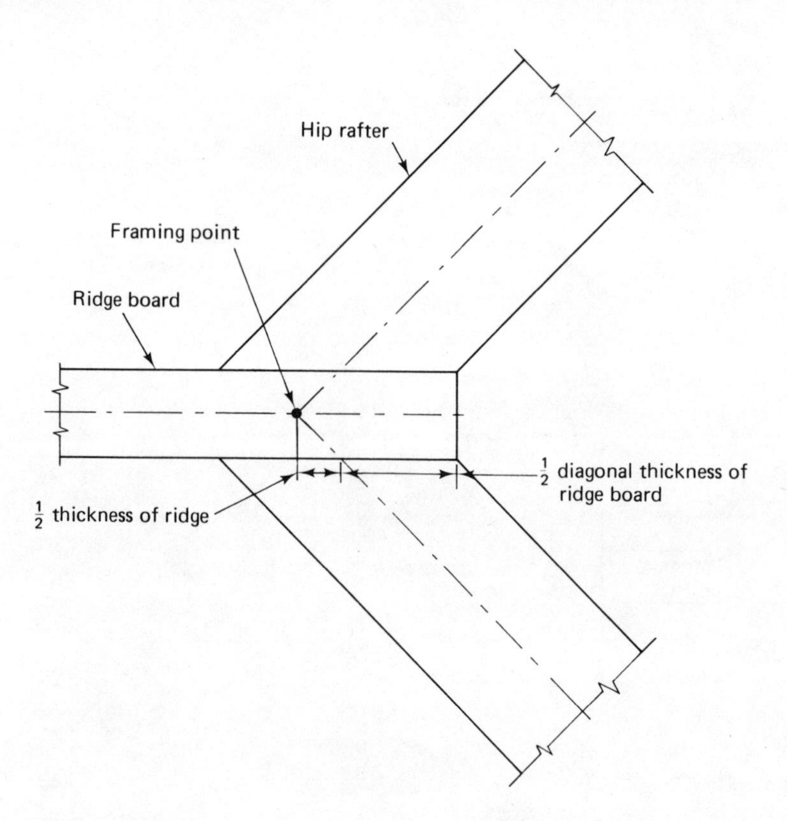

The diagonal thickness of a piece of framing stock is the thickness measured on a 45° line. Diagonal thickness of any piece of material can be determined quickly by drawing a line across the stock at a 45° angle and measuring the length of the line. The diagonal thicknesses for the more commonly used rafter stock are given in Table 3-1.

TABLE 3-1
Diagonal Thickness of Commonly Used Rafter Stock
(to the nearest $\frac{1}{16}''$)

Nominal Thickness	*Actual Thickness*	*Diagonal Thickness*
2″	$1\frac{1}{2}''$	$2\frac{1}{8}''$
1″	$\frac{3}{4}$	$1\frac{1}{16}''$
$\frac{1}{2}''$	$\frac{3}{8}''$	$\frac{9}{16}''$

Marking the Ridge Board

Layout marks on the ridge board should show the location of all the common rafters that frame against it. Also, the framing point for the hip rafters should be marked on the top edge of the ridge board, and the long point of the hip rafter should be shown if appropriate.

It is common practice to place common rafters directly above the studs. Therefore, the ridge board should be marked to maintain proper spacing and should locate the rafters in line with the studs. Various mechanics will develop their own system for marking the ridge board. One of the simplest methods is outlined in the following paragraphs.

The initial step in preparing to mark the ridge board is to measure in the run distance from each end of the building and mark it on the header joist (see Fig. 3-4). The distance between these marks is the theoretical ridge length. Since studs are placed

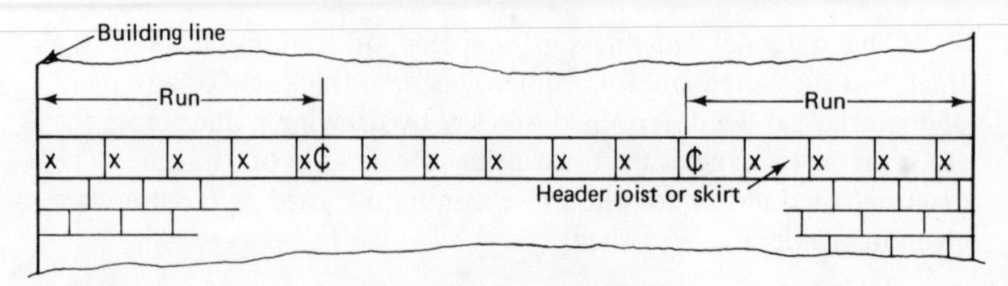

Locating Framing Points on Skirt

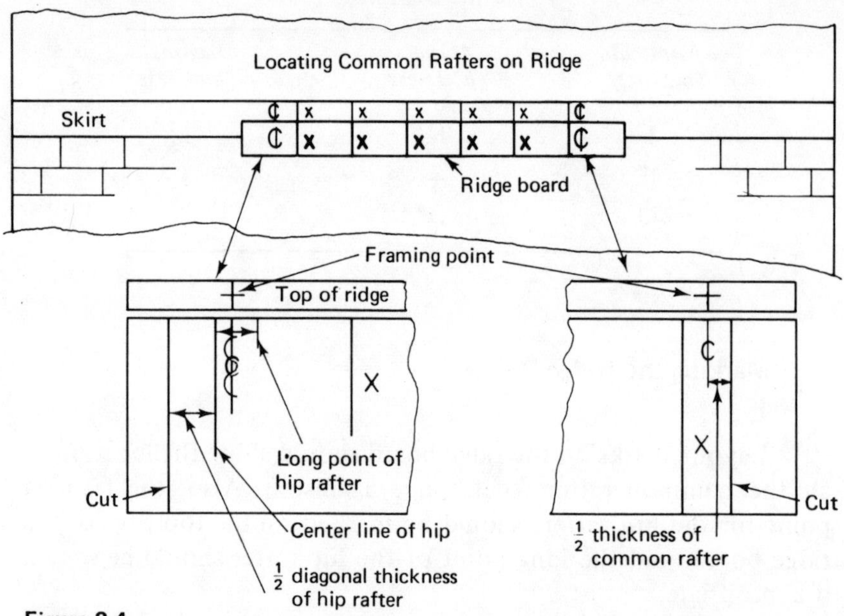

Figure 3-4
Marking the ridge board.

in line with the joists, the common rafter location can be marked on the ridge board by transferring joist marks from the header joist to the ridge board.

The framing point location is also marked on the top edge of the ridge board, and from the framing point the allowances can be marked out to locate the cutting line for ridge board length.

At each end where the hip rafters frame into the sides of three common rafters, the ridge is extended beyond the framing point one-half the thickness of the common rafter, but at each end where the hip rafters frame directly to the ridge board the ridge is extended by an amount equal to one-half the thickness of the ridge board plus one-half the diagonal thickness of the hip rafter.

RUN OF THE HIP RAFTER

The hip rafter moves into the building at a 45° angle in plan view. Therefore, its actual run is the diagonal of a square with sides equal to the run of the common rafter. It is seldom necessary to know the actual run of the hip rafter because the unit length of the hip rafter which is given on the rafter framing table is calculated for one unit of common rafter run.

By using the common rafter unit run (12″) for the length of the sides of the square of which the diagonal is the hip rafter unit run, the unit run of the hip rafter may be calculated. Working with two sides of this 12″ square and the diagonal, we have a 45° triangle with legs of 12″, and by using the Pythagorean theorem we find the length of the third side to be 16.97″. For practical purposes, in doing layout of plumb and level lines, 17″ is used as the unit run for all hip rafters and valley rafters in equal pitch roofs.

If it is necessary to determine the total run of the hip rafter, the unit run is multiplied by the number of units of common rafter run. Seventeen inches can be used as the unit of run in this problem, but for greater accuracy 16.97″ should be used as the unit of run.

EXAMPLE:

Common rafter run = 13′

Hip rafter run = 16.97 × 13 = 220.61″

Hip rafter run = 17 × 13 = 221″

The difference of .39″ in run amounts to a greater increase in the length of the rafter. Therefore, for accuracy in determining total run, 16.97″ should be used.

Figure 3-5 shows the relationship between the common rafter and hip rafter unit run, unit rise, and unit length. It should be noted that although common and hip rafters have different units of run, they have the same unit rise per unit of run. Because the hip rafter unit run is based on the common rafter unit run, hip rafters and common rafters in the same roof have the same number of units of run, rise, and length.

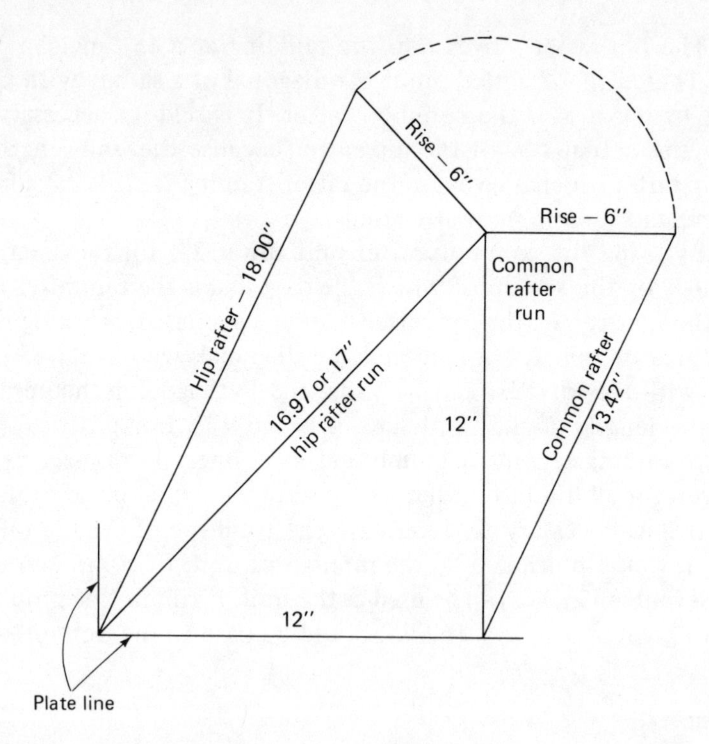

Figure 3-5
Relationship of hip rafter and common rafter.

HIP RAFTER LAYOUT

The length of the hip rafter may be scaled in the same manner as a common rafter (pages 20–22) when the total rise and total run are known. It may also be scaled by using the slant triangle method, and the slant triangle method may be used to calculate the length of the rafter using the Pythagorean theorem. Other methods for determining hip rafter length are the step-off method and the unit-length method.

Hip Rafter Length

The hip rafter has the same total rise as a common rafter in the same roof. A method for determining total run was discussed in the previous paragraph. Although this method is theoretically correct, the fact that the total run must be calculated usually makes scaling total rise and run unacceptable for determining hip rafter length.

The length of the hip rafter may be determined by scaling the diagonal distance of the common rafter length and the common rafter run. This is called the *slant triangle method* for determining hip rafter length because it involves taking a triangle that lies on the slope of the roof. The advantage of using this method is that the length of the common rafter and its run are already known. In this method of finding the length of the hip rafter, the length of the common rafter is held on the body of the square and the run of the common rafter is held on the tongue. The diagonal distance is marked on the work surface and measured (see Fig. 3-6). The run of the common rafter in Fig. 3-6 is 13′, and the length of the common rafter is 16′–3″. By holding 13″ on the tongue of the square and $16\frac{3}{12}''$ on the body of the square, the diagonal distance is scaled and found to be $20\frac{9.75}{12}''$. Therefore, the rafter length is $20'\text{–}9\frac{3}{4}''$.

The line length of the hip rafter may be calculated (using the slant triangle) by finding the square root of the sum of the square of the common rafter length and common rafter run.

EXAMPLE:

Common rafter length = 15′–0″

Common rafter run = 12′–0″

$$\text{Hip rafter length} = \sqrt{15^2 + 12^2}$$

$$\text{Hip rafter length} = \sqrt{225 + 144}$$

$$\text{Hip rafter length} = \sqrt{369} = 19.21' = 19'\text{–}2\tfrac{1}{2}''$$

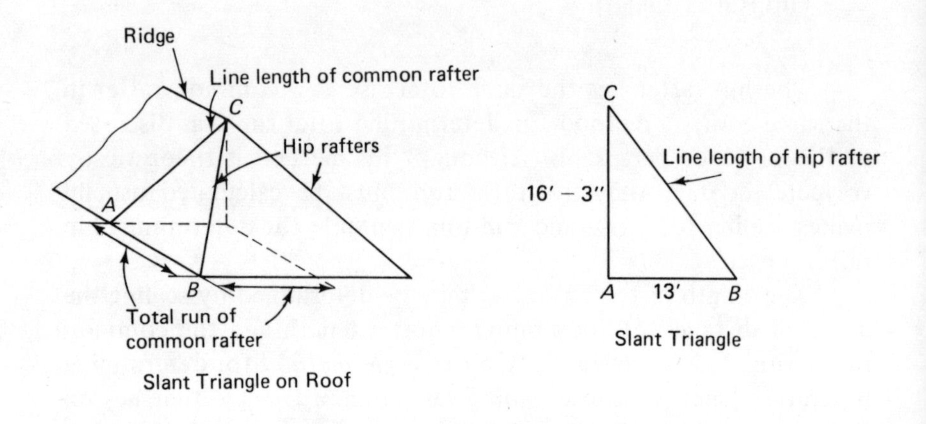

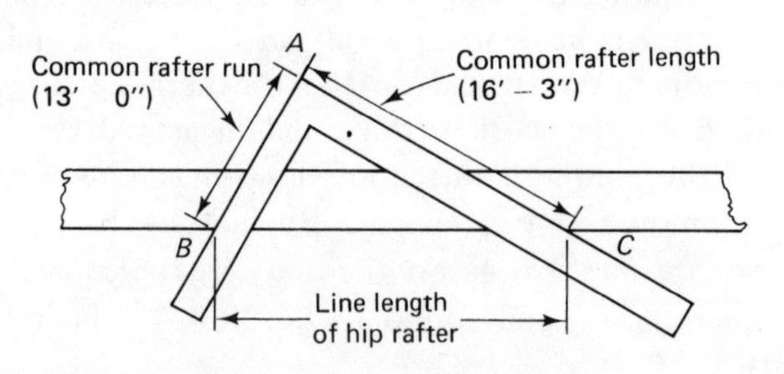

Figure 3-6
Slant triangle method for hip rafter length.

The unit-step method of determining hip rafter length is similar to that for determining the length of the common rafter. The only difference is that 17″ is used for the unit of run.

EXAMPLE:

Common rafter run	=	13′–4″
Unit rise	=	9″
Hip rafter unit run	=	17″

Since there are $13\frac{1}{3}$ units of run, the framing square is stepped 13 times and marked holding 9″ and 17″. A fourteenth step is made, but instead of marking at 17″, a mark is made at $5\frac{8}{12}$″ ($5\frac{11}{16}$″) because only one-third of the last unit was needed (see Fig. 3-7).

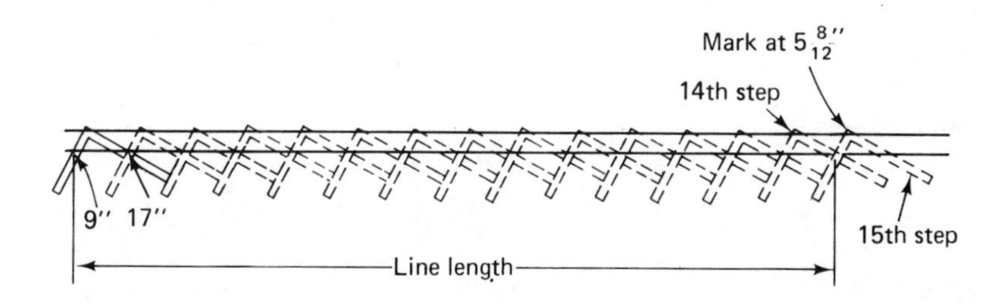

Figure 3-7
Step method of determining hip rafter length.

The common rafter run may work out in some odd fraction such as 13′–$7\frac{5}{8}$″. To step off the $7\frac{5}{8}$″ for a common rafter is a simple matter, but for a hip rafter the diagonal of $7\frac{5}{8}$″ must be found. The diagonal of $7\frac{5}{8}$″ is the part of 17″ that would have to be stepped off in this case. The simplest way to determine this distance is to hold $7\frac{5}{8}$″ on each side of the square and measure the diagonal, $10\frac{3}{4}$″ (see Fig. 3-8).

Perhaps the most commonly used method for determining hip rafter length is the unit method employing the table on the rafter framing square. The unit length of the hip rafter is given on the second line of the rafter framing square. The line length of the

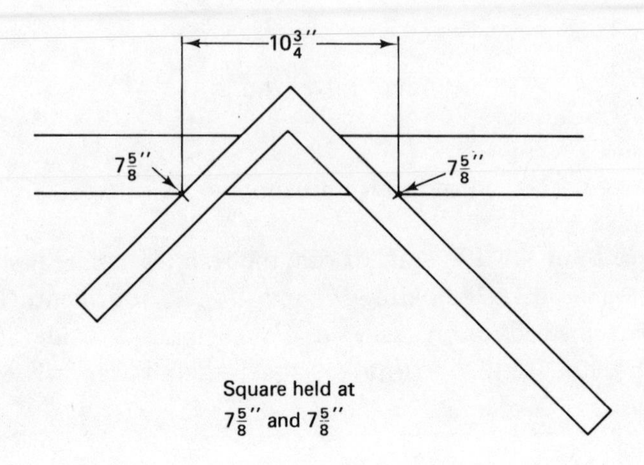

Figure 3-8
Finding proportioned
Part of run.

hip rafter is determined by multiplying the unit length for a given unit rise by the number of units of run covered by the common rafter.

EXAMPLE:

Run of common rafter = 13′ Unit rise = 4″

Unit length of hip rafter = 17.44″

$$17.44 \times 13 \text{ units} = 226.72''$$

$$0.72'' \times \frac{16}{16} = \frac{11.52}{16} \text{ or } \frac{3}{4}''$$

$$226'' \div 12'' = 18'\text{–}10''$$

$$\text{Line length} = 18'\text{–}10\tfrac{3}{4}''$$

This mathematical line length is transferred to the rafter stock and is laid out on the top edge of the stock in a manner similar to that for common rafters (see Fig. 3-9). One line is squared across the stock near the upper end of the rafter and another line is squared across the stock at the line-length mark. Plumb lines are drawn on the side of the rafter stock at the point marked by the lines drawn on the top edge of the stock.

Plumb lines for the hip rafter are laid out by holding the unit rise of the roof on the tongue of the square and the unit run for hip rafters, 17″, on the body of the square and marking on the rise side of the square. To lay out level lines, the mark is drawn on the run side of the square.

Figure 3-9
Marking line length on hip rafter stock.

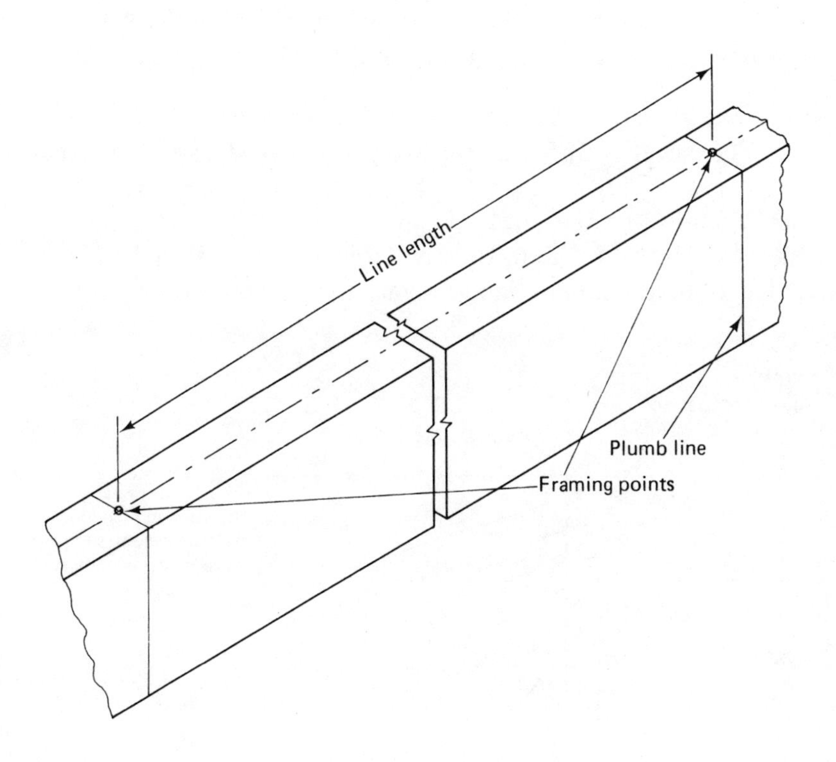

Bevel or Side Cuts

The bevel or side cut on hips, valleys, and jacks may be laid out by using the side cut table on the last two lines of the framing square. When this table is used, the number given under the unit rise is held on one side of the square and 12″ is held on the other. The bevel is always marked on the 12″ side (see Fig. 3-10). It is necessary to mark the shape of the bevel when rafters are cut by hand.

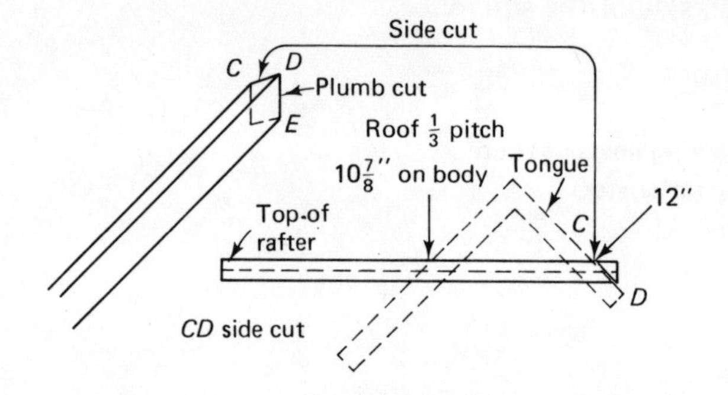

Figure 3-10
Marking bevels.

The bevel is different for every pitch of roof because it is a compound angle. In plan view, however, the bevels appear as 45° cuts regardless of the pitch of the roof (see Fig. 3-11). When portable power saws are used, the saw table is set to 45° and the saw is guided along the plumb line to create the proper bevel.

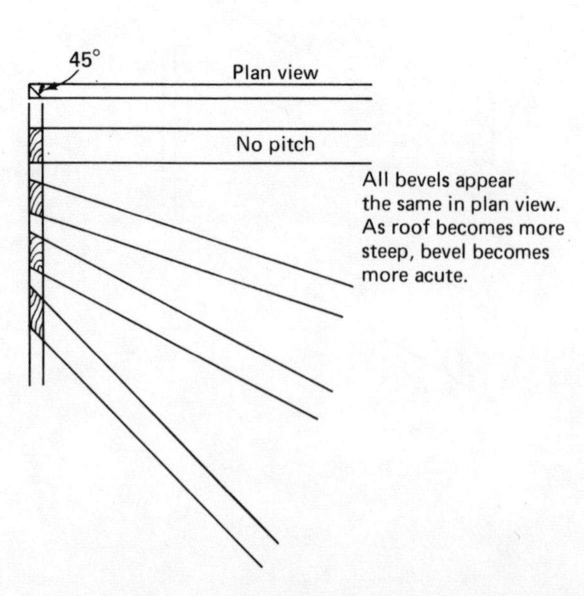

Figure 3-11
Bevel comparison.

Hip Rafter Shortening and Bevels

An allowance for the ridge board must be made in determining the actual length of the hip rafter. The amount allowed depends on the method by which the hip rafter is framed. When hip rafters are framed directly to the ridge board, they meet the ridge as shown in plan view in Fig. 3-12. The line length is figured to the framing point, which is at the center of the ridge board. Therefore, the rafter must be shortened an amount equal to one-half the diagonal thickness of the ridge board. Because this is a plan-view distance, it must be deducted by measuring at right angles to the plumb line.

Figure 3-12
Hip rafter framing to ridge — plan view.

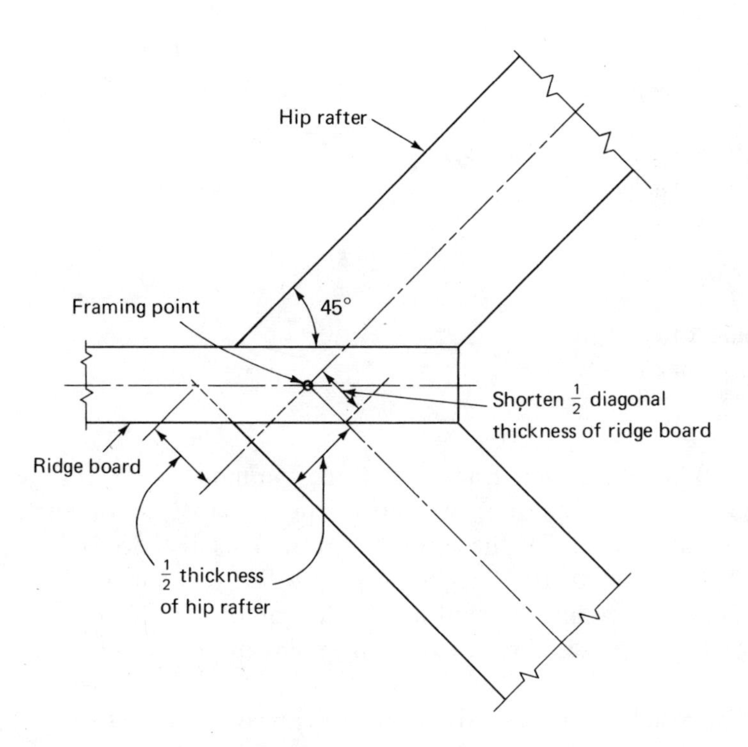

To locate the long point of the bevel a distance equal to one-half, the thickness of the rafter stock is marked off as shown in Fig. 3-13.

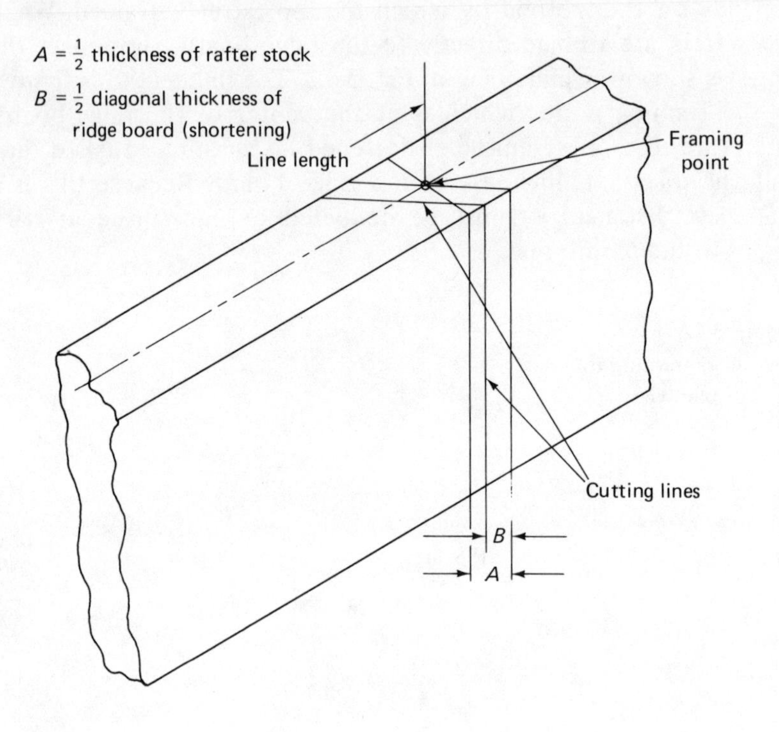

Figure 3-13
Layout of hip rafter bevel — framing to ridge.

When hip rafters frame into three common rafters as shown in Fig. 3-14, the amount of shortening is equal to one-half the diagonal thickness of the common rafter. This distance is deducted at right angles to the line-length plumb line and a new plumb line is drawn in. A line squared across the top of the rafter at the new plumb line passes through the long point of the double half-bevel of the rafter. To locate the short point of this half-bevel, a distance equal to one-half the rafter thickness is marked off at right

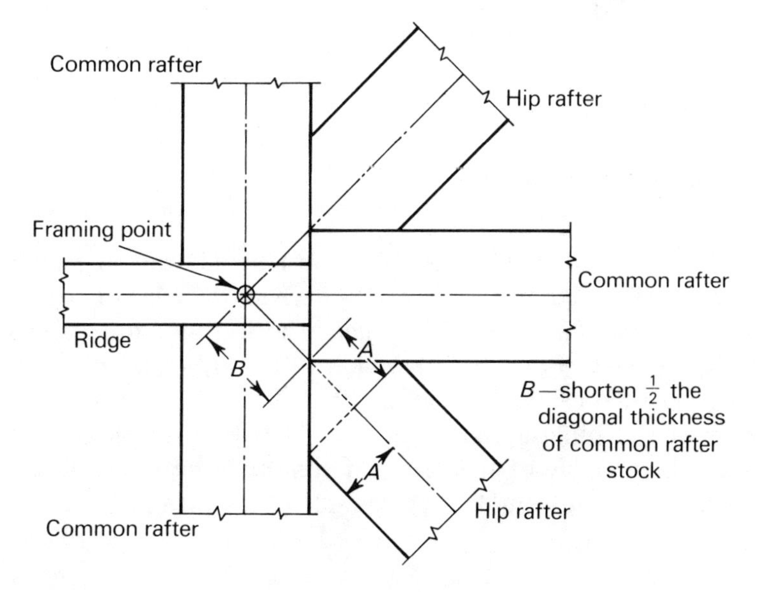

Figure 3-14
Hip rafter framing to common rafters — plan view.

angles to the shortening plumb line and a third plumb line is drawn on the side of the rafter. The shape of the bevel may be drawn on the top edge of the rafter stock by connecting the third plumb line with the long point of the double half-bevel located at the center of the rafter on the shortening line (see Fig. 3-15).

Figure 3-15
Layout of hip rafter bevels — framing to common rafters.

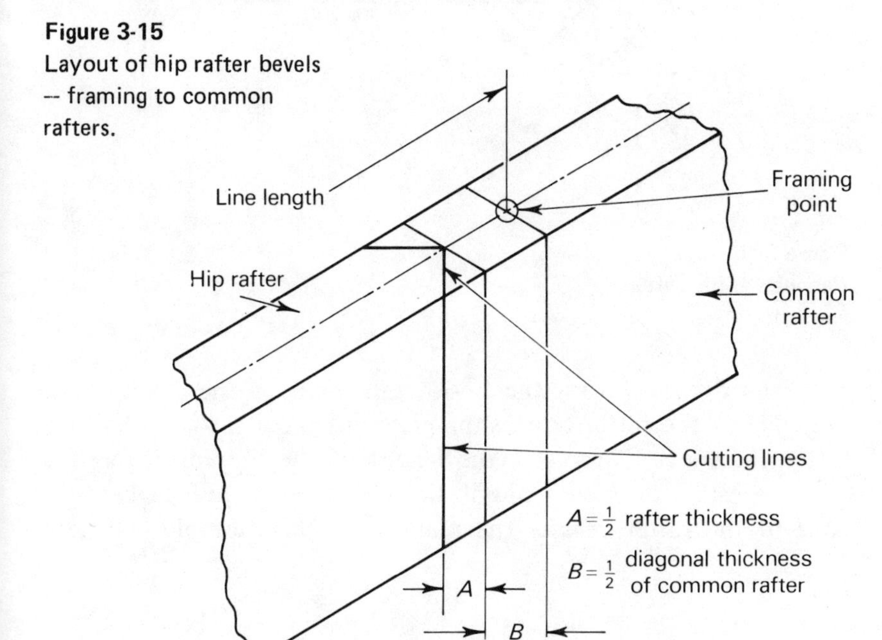

Hip Rafter Seat Cut

The seat cut of the hip rafter must be laid out in a manner that will make its height at the point where the edge of the rafter crosses the plate line equal to the height of the common rafter seat cut. Figure 3-16 shows the plan view of the hip rafter as it crosses the corner of the building. Notice that the line length falls at the corner of the building and that when the line length is squared across the top edge of the rafter, it is away from the plate line at the point where it is marked on the side of the rafter.

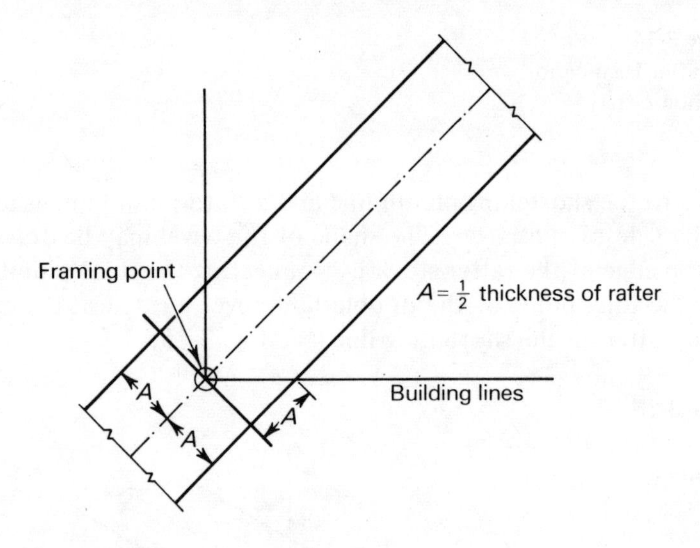

Figure 3-16
Plan view — hip rafter seat cut.

The distance from the line-length mark to the point where the edge of the rafter crosses the plate line is equal to one-half the rafter thickness. Therefore, the height of the hip rafter seat cut must be established on a plumb line drawn at the point where the edge of the rafter crosses the plate line. This new plumb line is located by measuring in at right angles to the plumb line a distance equal to one-half the rafter thickness. The height of the common rafter seat cut is established on the second plumb line and the level line is drawn in as illustrated in Fig. 3-17.

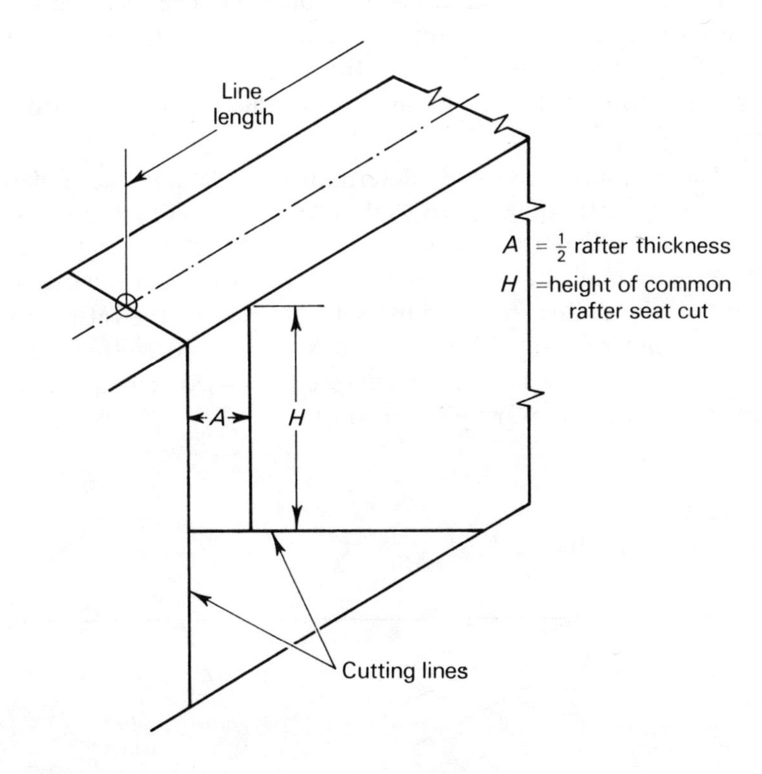

Figure 3-17
Layout of hip rafter seat cut.

Dropping and Backing the Hip Rafter

An alternate method for laying out hip rafter seat cuts requires that the hip rafter either be dropped or be backed. *Dropping* is a process in which an additional amount of material is removed from the seat cut. *Backing* is a process in which the top edge of the hip rafter is beveled to allow the roof sheathing to be fully supported on the hip rafter.

When the height of the common rafter seat cut is marked on the line-length mark, the height of the common rafter is established at the framing point and along the center line of the entire rafter length. Since this procedure causes the edges of the hip rafter to be above the top of the common rafter, the hip rafter must be dropped.

The amount of drop is determined by holding the unit rise and 17″ on the framing square and measuring in along the run side of the square a distance of one-half the rafter thickness and making a mark. A plumb line is drawn through the mark and the distance along the plumb line from the mark to the edge of the rafter stock is the amount of drop. This distance is transferred to the seat cut layout and a new level line is drawn (see Fig. 3-18). The amount of drop is different for every pitch of roof.

Figure 3-18
Dropping the hip rafter.

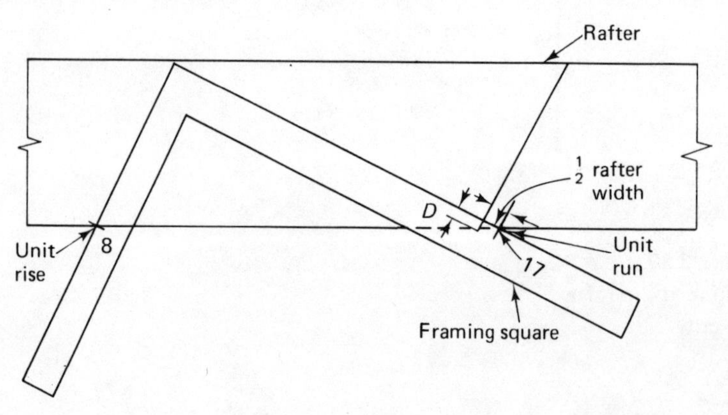

1. Hold square so that the distance from the edge of rafter to length plumb line is equal to $\frac{1}{2}$ rafter thickness.
2. Place mark at $\frac{1}{2}$ rafter thickness.
3. Measure along plumb line the distance from the top edge of the rafter to mark.
4. Mark distance *D* on seat cut.

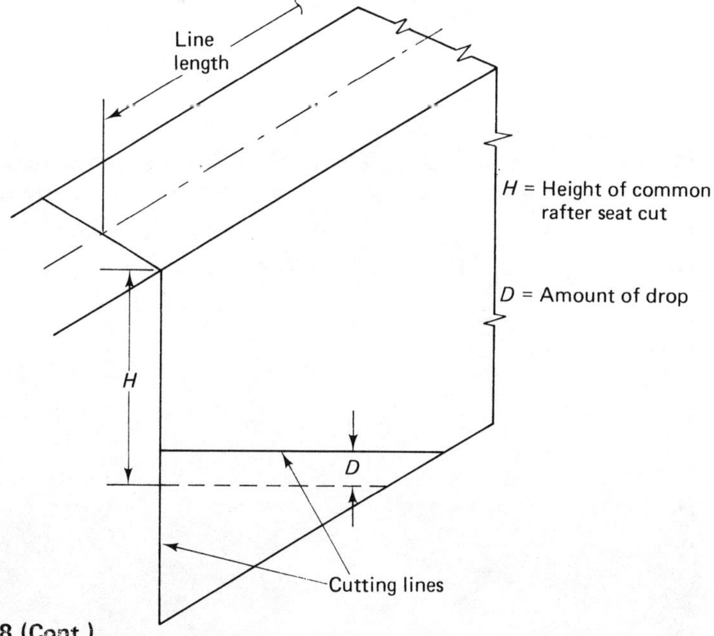

Figure 3-18 (Cont.)

The hip rafter is not dropped when it is backed. The height of the common rafter seat cut is marked on the line length mark. The amount of backing is determined by holding the unit rise and 17″ on the framing square and measuring in along the run side of the square and making a mark at a distance of one-half the rafter thickness (see Fig. 3-19). A line is drawn through this mark parallel

Figure 3-19
Determining the amount of backing.

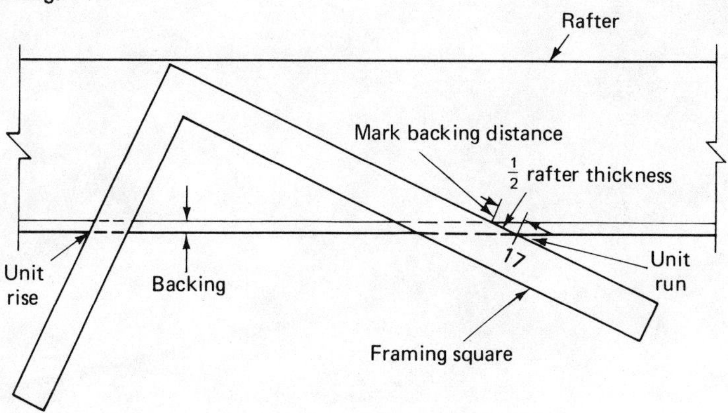

to the edge of the rafter stock for the entire length of the rafter. A line is drawn on the other side of the rafter the same distance from the top edge. In cutting the backing, the power saw is set so that it follows the line on the side of the rafter stock and comes out at the center of the top edge (see Fig. 3-20).

Figure 3-20
Backing the hip rafter.

Tail Layout

The tail of the hip rafter has the same number of units of run as the common rafter tail in the same roof, but the actual run of the hip rafter tail is equal to the diagonal of the cornice projection and may be determined by multiplying the number of units of cornice projection by 16.97″. To determine the length of the tail, either the rise and run may be stepped off or the unit length of the hip rafter may be multiplied by the number of units of cornice projection.

The line length of the tail is marked on the top edge of the hip rafter, starting at the line-length mark at the seat cut as shown in Fig. 3-21. The bevels at the cornice line are laid out so that

Figure 3-21
Layout of hip rafter tail.

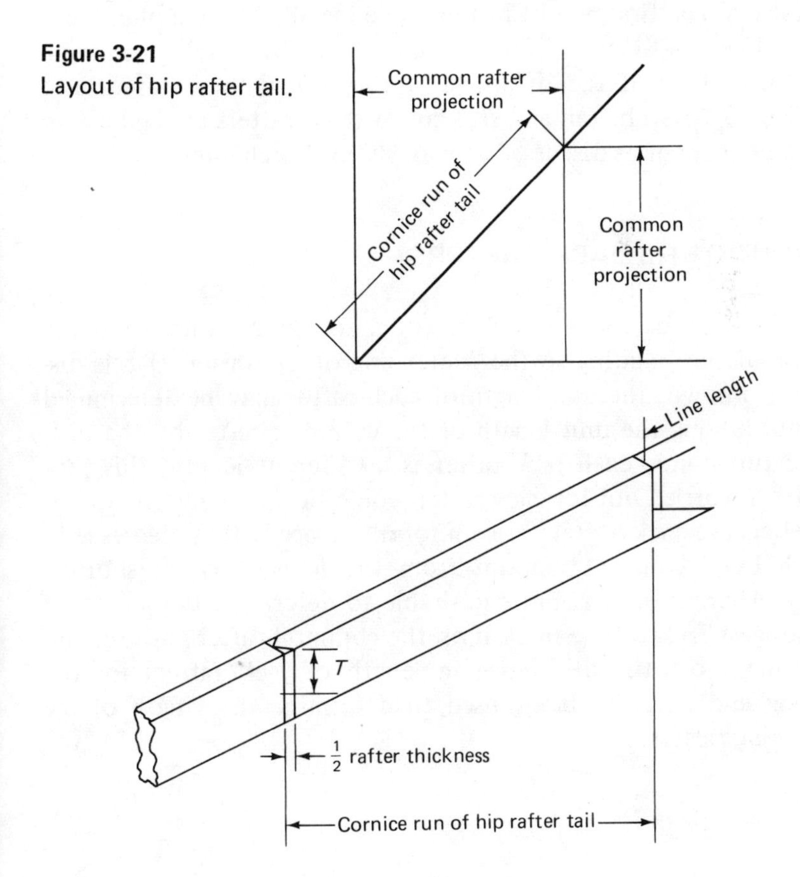

the long point of the tail falls at the length of the tail mark. The horizontal line at the end of the tail is located by transferring the height of the common rafter tail to the cutting lines of the double half-bevel at the tail. Marking at this point results in a hip rafter tail which has the same height as the common rafter tail at the cornice line.

HIP RAFTER LAYOUT FOR PYRAMID ROOFS

Hip rafters for a pyramid roof meet at a common point over the center of the building. The line length of the hip rafters is determined and marked on the rafter stock as previously discussed.

It is not necessary to make bevel cuts at the top end of the hip rafters. The first pair of rafters to be installed have a plumb cut at the line-length mark. The second pair to be installed must be shortened one-half the thickness of the rafters they frame onto (see Fig. 3-22). The tail and seat cut of these rafters are laid out in the same manner as discussed previously in this chapter.

LAYOUT OF HIP JACK RAFTERS

The run of the hip jack is equal to the distance from the corner of the building to the center line of the rafter. If this distance is known, the line length of each rafter may be determined by multiplying the unit length of the common rafter by the jack rafter run. Since each jack rafter is a different length, this procedure, if carried out for each rafter, could be a lengthy process.

Because jack rafters are uniformly spaced, they decrease in length by a constant amount from the longest to the shortest rafter. Therefore, it is more desirable to determine the length of the longest jack rafter, mark it on the common rafter pattern, and then mark off the difference in length of jack rafters for the spacing and unit rise being used to determine the length of the remaining rafters.

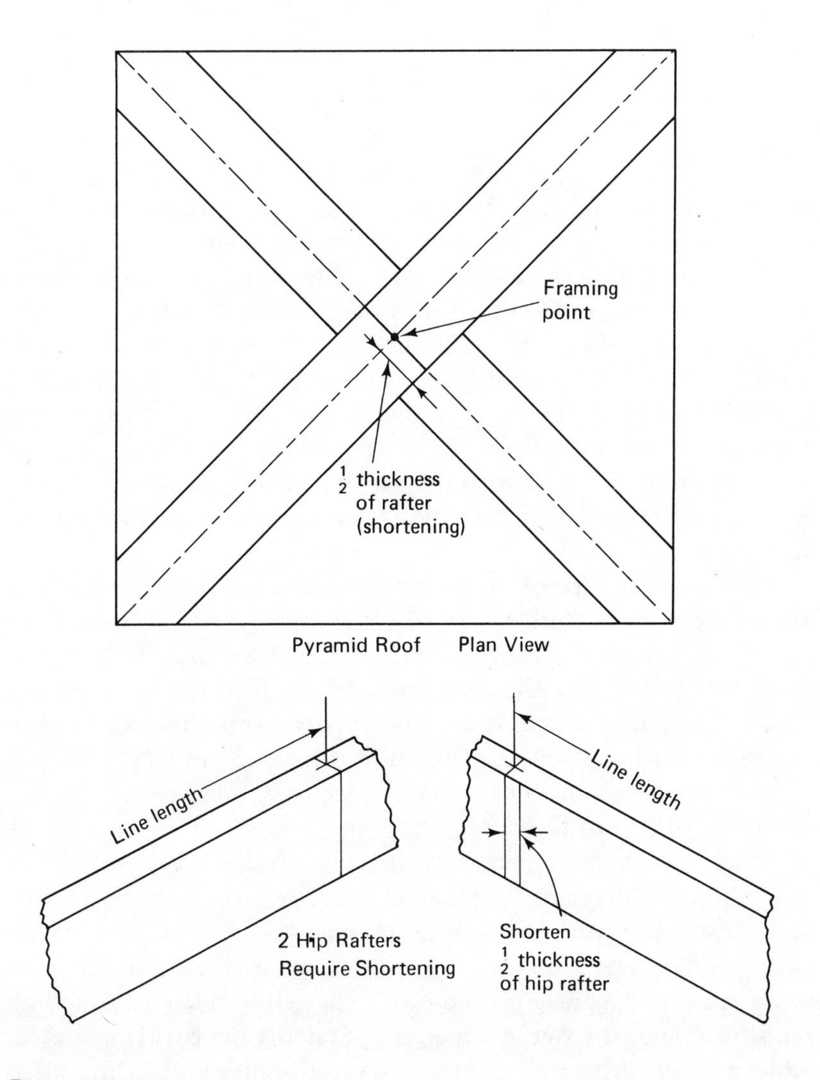

Figure 3-22
Pyramid roof.

The difference in length of jack rafters spaced 16″ O.C. is given on the third line of the rafter framing table under the various unit rises on the framing square. It was determined by multiplying the unit length of the common rafter by $1\frac{1}{3}$ because 16″ is $1\frac{1}{3}$ units of run. The difference in the length of jack rafters spaced 24″ O.C.

is found on the fourth line of the rafter table. It was determined by multiplying the common rafter unit length by 2, because 24″ is two units of run. For jack rafters 12″ O.C., the difference in length is the same as the common rafter unit length.

The length of the longest jack rafter may be found by determining where it is placed in the roof in relation to the hip rafter and the ridge board. The simplest way to determine this is to measure on the ridge board the distance from the far side of the common rafter to the location where the long point of the hip rafter bevel will join the ridge. This distance is called *C* in Fig. 3-23. By subtracting *C* from the rafter spacing, distance *D*, which is the amount of run lost by the longest jack rafter, can be determined.

The length of the longest jack rafter at the long point is determined by measuring distance *D* at right angles to the plumb line of the common rafter pattern. A plumb line may be drawn at this point to transfer the long point of the jack rafter to the top edge of the rafter stock. It is then squared across the top edge of the rafter pattern, and the common difference in length of jack rafters for the given unit rise and spacing is marked off for the remaining length of the rafter.

It is usually not necessary to lay out the bevels for a jack rafter because they are automatically cut by setting the power saw table at 45° and following the plumb line. If jack rafters are cut by hand or if it is necessary to have the shape of the bevel for some other reason, they may be laid out by using the side cuts found on the fifth line of the steel square rafter framing table. If the side cut table is used, the number given under the unit rise of the roof is held on one side of the square, 12″ is held on the other side, and the shape of the bevel is marked along the 12″ side.

Bevels or side cuts for jack rafters may also be laid out by measuring back at right angles to the plumb line an amount equal to the rafter thickness and drawing a second plumb line. The length at the second plumb line is squared across the top of the rafter stock, and a diagonal line connecting the two lines squared across the top of the stock represents the shape of the bevel or side cut.

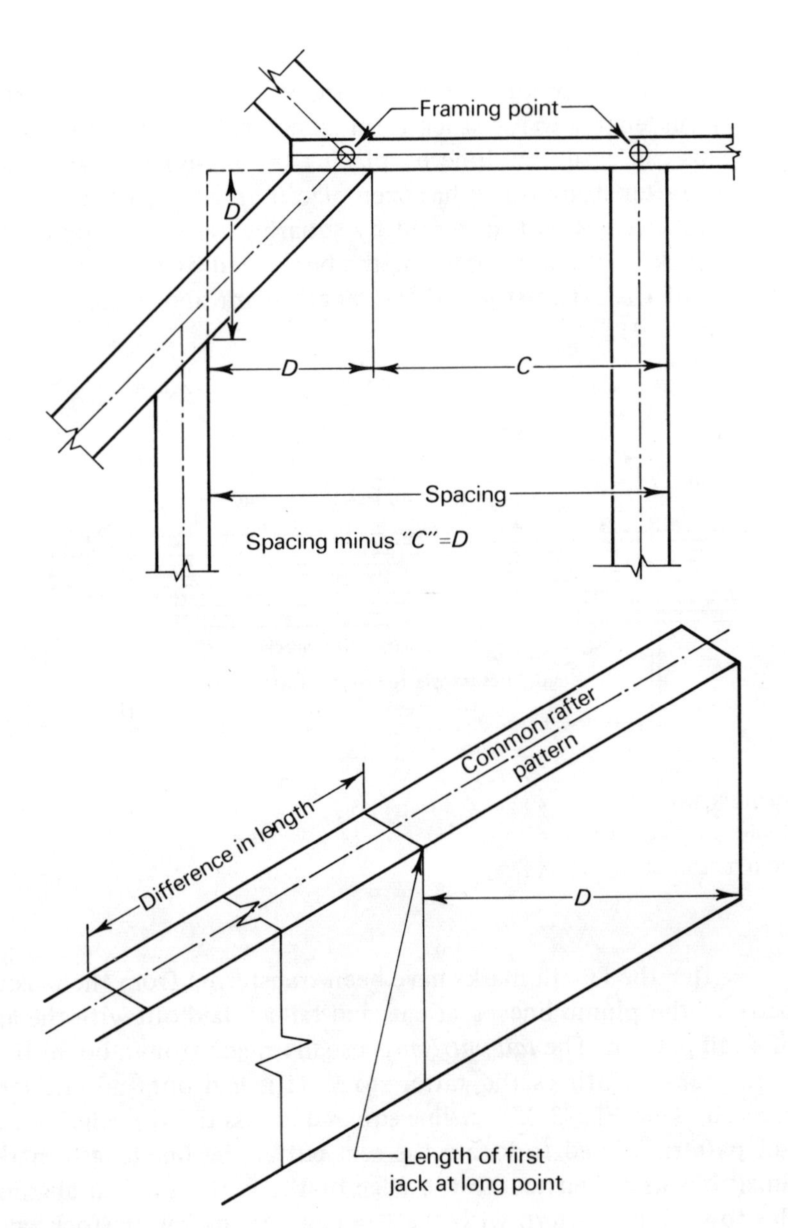

Figure 3-23
Layout of hip jack rafter master pattern.

After the master pattern for the jack rafters has been completed on the common rafter pattern, it is necessary to transfer the various lengths to the stock form from which the rafters will be cut. This can easily be done by placing the master pattern alongside the rafter stock which has been placed on edge on a set of saw horses. The length is transferred by squaring across the top of the rafter stock at the long point of the bevels and by squaring across the top of the rafter stock at the length mark above the seat cut (see Fig. 3-24).

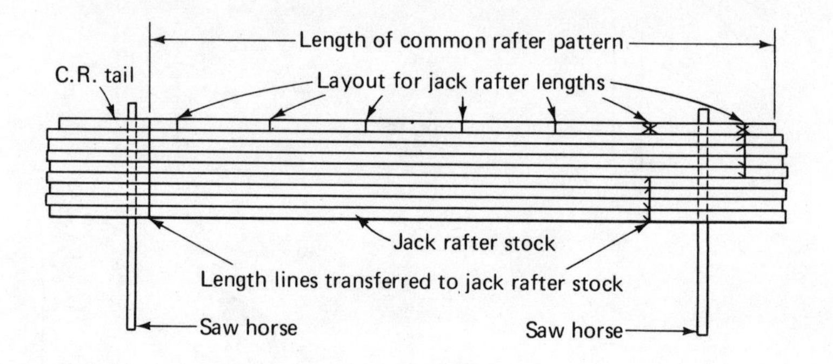

Figure 3-24
Transfer of layout marks from master pattern.

After the length marks have been transferred from the master pattern, the plumb lines, seat cut, and tail are laid out with the aid of a tail pattern. The *tail pattern* is usually made from a board that is the same width as the rafter stock. It is laid out and cut very carefully (see Fig. 3-25). A line squared across the top edge of the tail pattern is used to locate the seat cut at the line-length mark. Small blocks nailed to the top edge of the pattern aid in aligning the top of the pattern with the top edge of the rafter stock, and the plumb cut at the end of the pattern can be used to draw plumb lines where needed on the jack rafter stock.

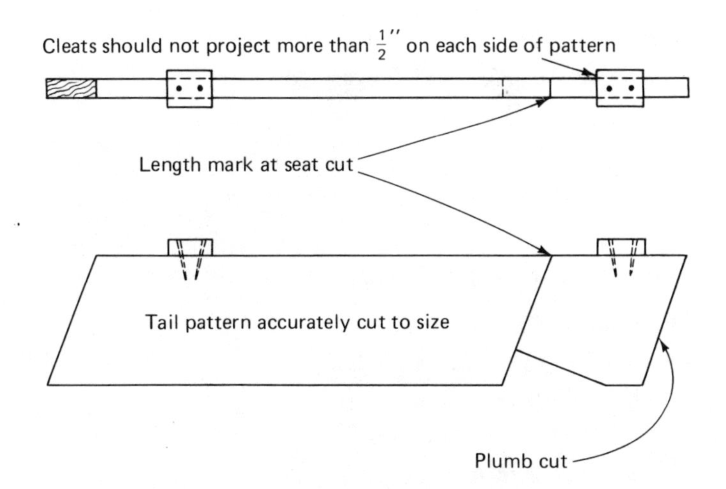

Figure 3-25
Tail Pattern

REVIEW QUESTIONS

1. Where are hip rafters placed? At what angle are they placed?
2. Why are hip rafters made from wider stock?
3. When is it necessary to make a roof plan?
4. Sketch a layout for the ridge board in a hip roof.
5. What is diagonal thickness?
6. What is the diagonal thickness of $1\frac{1}{2}''$? Of 3/4"? Of 3/8"?
7. What is the actual run of a hip rafter?
8. What is the unit run of a hip rafter?
9. Do hip rafters and common rafters in the same roof have the same rise?
10. Do hip and common rafters in the same roof have the same number of units of run?
11. How can the length of the common rafter be used to calculate the length of a hip rafter?
12. What is a slant triangle?

13. Where may the unit length of hip rafters be determined?
14. Calculate the length of the following hip rafters:

Unit Rise	*Run*
5″	13′
6″	12′–8″
9″	13′–7″

15. How are bevel or side cuts laid out?
16. What is shortening?
17. Sketch the plan view and hip rafter layouts for hip rafters framing to a 3/4″ thick ridge board.
18. Sketch the plan view and hip rafter layouts for hip rafters framing to three common rafters.
19. Why is it necessary to drop or back a hip rafter?
20. Sketch the plan view and rafter layout for a hip rafter seat cut and tail.
21. Sketch the plan-view layout at the "ridge" of a pyramid roof.
22. How can the run of a hip jack rafter be determined?
23. How is the common difference in the length of jack rafters determined?
24. How is the master pattern for jack rafters laid out?

FOUR

Equal Pitch Intersecting Roofs

Buildings that have offsets or wings have intersecting roofs. They may be intersecting hip roofs, intersecting gable roofs, or a combination of hip and gable roofs. Equal pitch intersecting roofs fall into two categories: One is an intersecting roof in which the wing or bay is the same width as the main roof. The other is an intersecting roof in which the wing is narrower than the main part of the building.

ROOF PLAN

When a building has only one intersecting roof, it may not be necessary to have a roof plan, but as the number of intersecting roofs on a building increases, the need for a roof plan becomes apparent. The roof plan shows the location of the rafters in a level plane. As with the hip roof, the roof plan normally indicates the outline of the building wall, the cornice line, valley rafters, and ridge boards.

The roof plan in Fig. 4-1 illustrates two typical methods of building intersecting roofs. One wing of the building has the same span as the main building. The roof on this section has the same total rise as the main roof. Therefore, the ridge boards meet at a common height. The other wing is narrower than the main building. The smaller span has less total rise than the main roof.

A building that has many offsets and changes in span requires a roof plan. One building of this kind is illustrated in Fig. 4-2. Notice that a number of different roofs are possible. Without a roof plan the carpenter could become confused and build the wrong kind of roof.

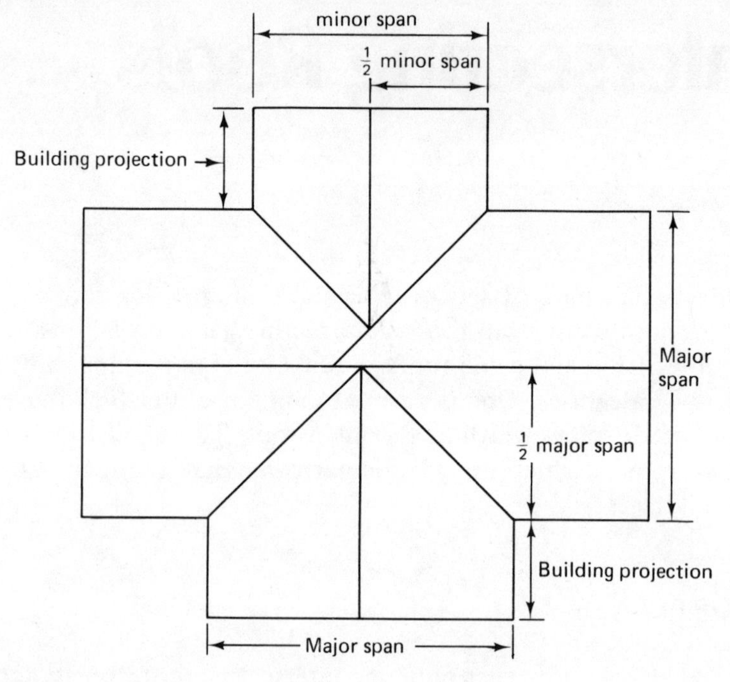

Figure 4-1
Roof plan.

EQUAL SPAN INTERSECTING ROOFS

The ridges in intersecting roofs having the same total span meet at a common height. The theoretical length of the ridge for roofs of this kind is equal to the run of the common rafter plus the length of the wing (see Fig. 4-1). The actual length is determined by shortening the ridge by an amount equal to one-half the thickness of the ridge board.

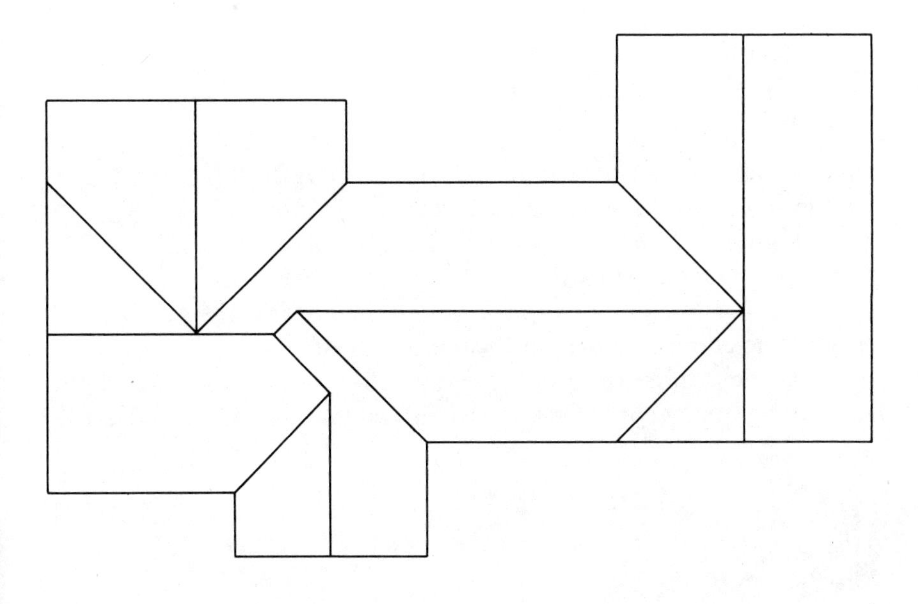

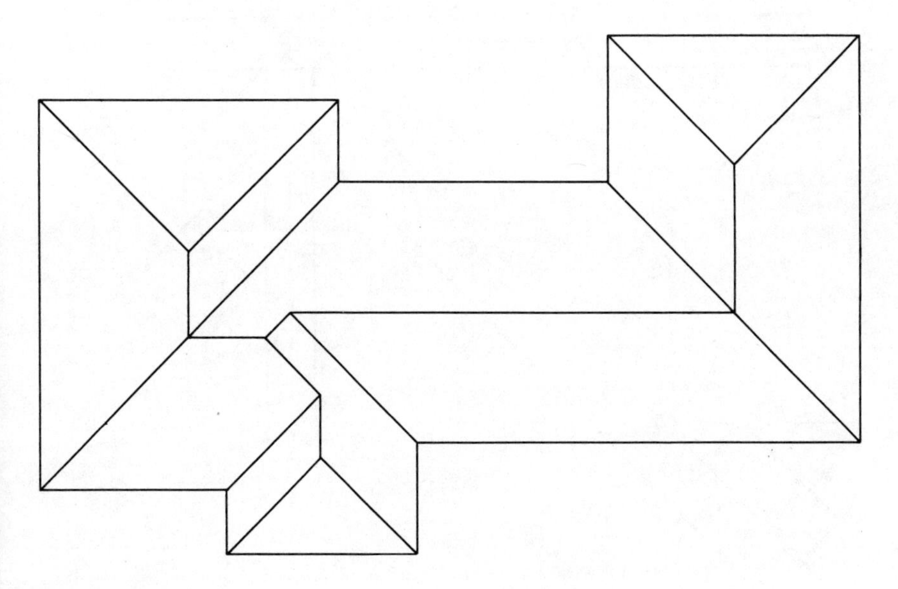

Figure 4-2
Complex roof plan.

Length of the Valley Rafter

The length of the valley rafter is determined in the same manner as the length of a hip rafter. Valley rafters and hip rafters in the same roof have the same theoretical length.

Figure 4-3 shows an enlarged plan view of the valley rafter framing to the ridge boards in an intersecting roof with equal spans. The number of units of run for this valley rafter is the same as the run of the common rafter in this kind of roof.

To determine the length of the valley rafter, the unit length found on the second line of the rafter framing table under the

Figure 4-3
Valley rafter
framing – plan view.

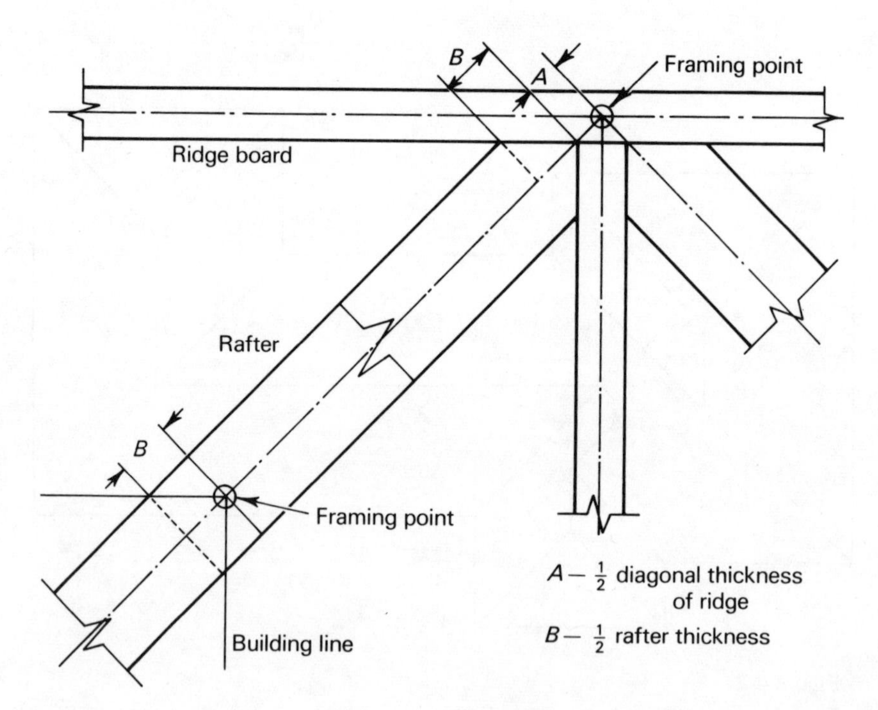

corresponding unit rise on the framing square is multiplied by the number of units of run.

EXAMPLE:

Unit rise = 8″

Span = 26′

Run = 13′

Unit length of valley rafter = 18.76″

Line length of valley rafter = 18.76 × 13 = 20′–3$\frac{7}{8}$″

The line length of the valley rafter is marked off on the top edge of the valley rafter stock in the same manner as for common rafters and hip rafters, and plumb lines are drawn on the side of the rafter to indicate the line length. Plumb lines for valley rafters are drawn by holding 17″ for the unit run on the body of the square and the unit rise on the tongue of the square and marking along the tongue or rise side of the square. Additional layout at the upper end must be made for shortening (to allow for ridge thickness) and for the shape of the bevels.

Valley Rafter Shortening and Bevels

The amount of shortening is equal to one-half the diagonal thickness of the ridge board and is deducted at right angles to the length plumb line at the upper end of the rafter (see Figs. 4-3 and 4-4). To locate the short point of the double half-bevel, an additional plumb line is drawn at a distance equal to one-half the rafter thickness from the shortening plumb line.

If the rafters are cut by a power saw, it is not necessary to draw the shape of the bevels on the top edge of the rafter. If the saw follows the plumb line with the table set at 45°, the proper bevels are automatically cut. If the cutting is done by hand, it is necessary to mark the bevels in order to provide a guide for the saw.

Valley Rafter Seat Cut

The line length for the valley rafter is calculated to the framing point which is located at the intersection of the two walls. When the line length is squared across the top of the rafter stock in plan view, it locates the plumb line at the side of the rafter stock and places it on top of the wall. Therefore, when the seat cut is laid out, it must be enlarged by an amount equal to one-half the rafter thickness measured at right angles to the plumb line.

The height of the seat cut is established by measuring from the top of the rafter along the length plumb line. The distance from the top edge of the rafter to the level line is equal to the height of the common rafter seat cut [see Fig. 4-4 (b)].

The layout for the tail may be accomplished by calculating the length using the run of the cornice and the unit length of the valley rafter. The complete layout of the tail for the valley rafter is illustrated in Fig. 4-4 (c). On many construction jobs the tail is not laid out completely. Instead, the tails of the common rafters are allowed to project to the corner of the building and are used to support the roof sheathing.

Figure 4-4 (a)

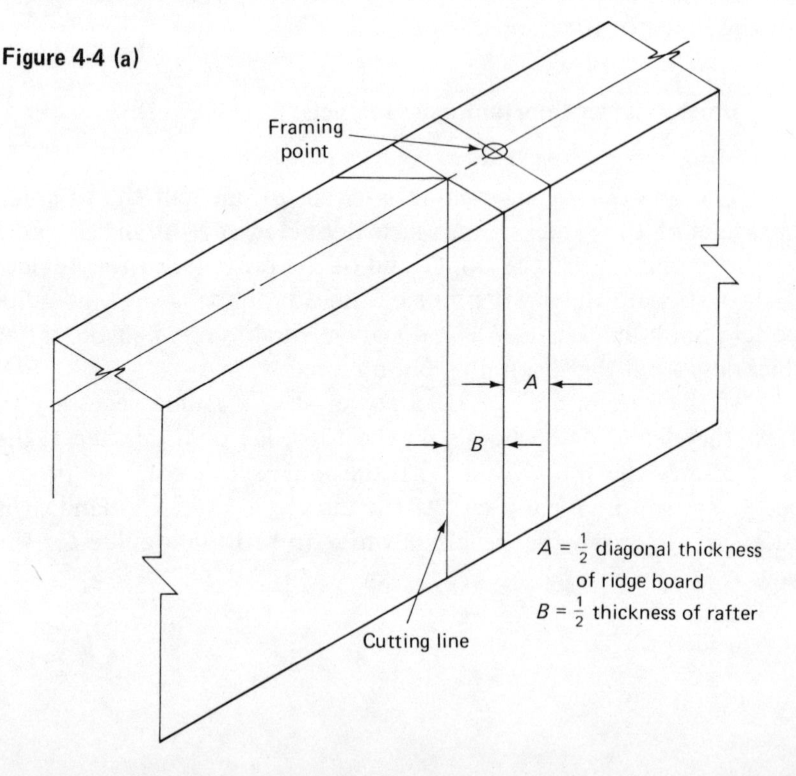

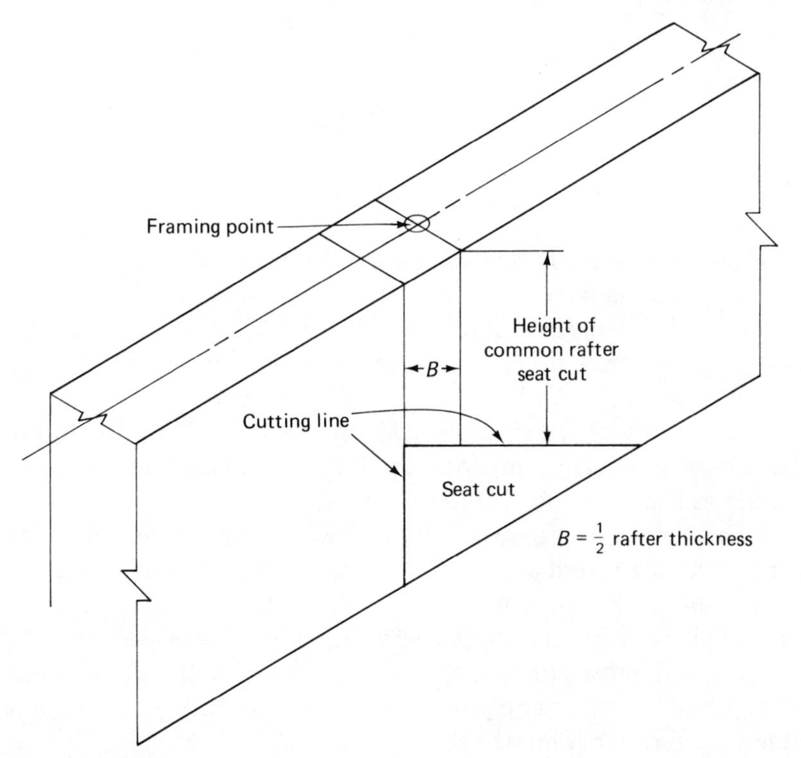

Figure 4-4 (b)
Valley rafter layout.

Figure 4-4 (c)
Valley rafter tail.

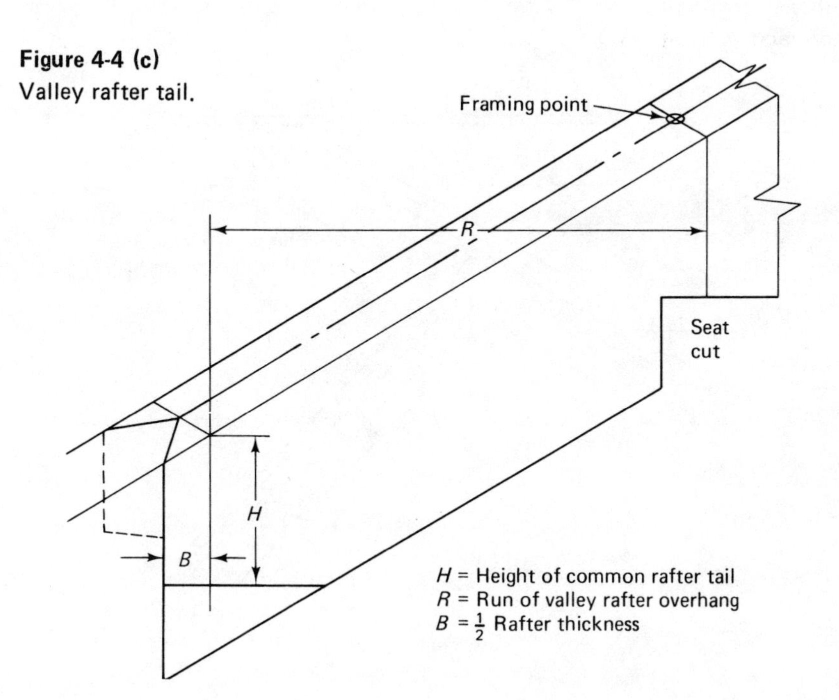

UNEQUAL SPAN INTERSECTING ROOFS

Intersecting roofs that have a smaller span than the main roof require a long supporting valley rafter and a short valley rafter (see Fig. 4-5). The length of the long valley rafter is governed by the run of the major span and is marked on the valley rafter stock in the same manner as for hip or other valley rafters.

The length of the short valley rafter is governed by the run of the minor span. It is marked on the rafter stock in the same manner as the length for other valley rafters.

The length of these valley rafters may be determined by scaling the diagonal distance of total rise and total run. It may be determined by stepping off in the 1/12 scale, and it may also be calculated by using the Pythagorean theorem (see pages 10 – 11). However, the most commonly used method of determining valley rafter length involves applying the unit method using the framing table on the rafter framing square.

Figure 4-5
Intersecting roofs of unequal spans.

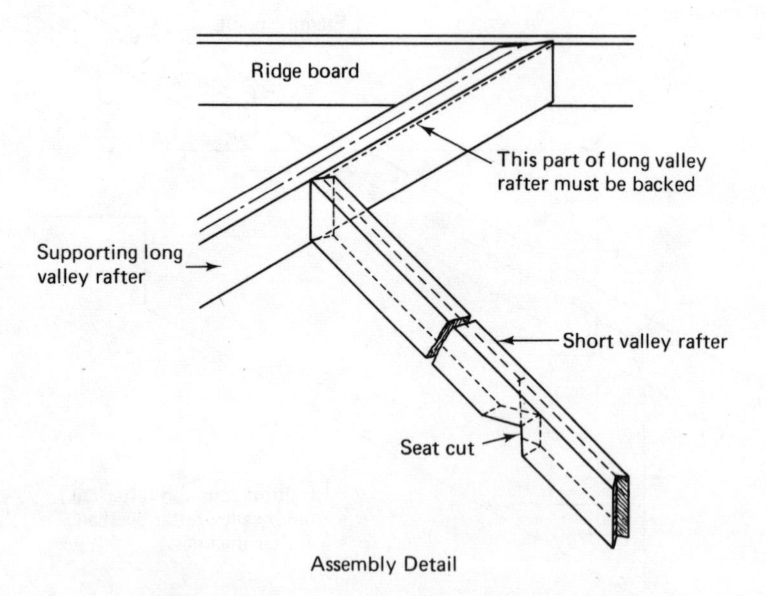

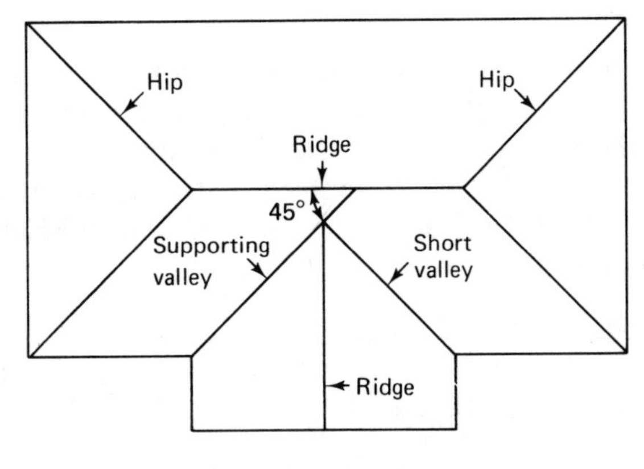

Figure 4-5 (Cont.)

Rafter Length

Valley rafters have the same unit of rise as common rafters in the same roof. The valley rafter unit of run is 17″. To determine the line length of a valley rafter, the unit length of the valley rafter is multiplied by the number of units of run it covers. The unit length is located below the unit rise on the second line of the rafter framing table.

EXAMPLE:

Major span = 28′ Unit rise = 10″

Minor span = 22′ Unit length = 19.70″

Length of long valley = 19.70 × 14 = 275.80″

$= 22'\text{–}10\frac{13}{16}''$

Length of short valley = 19.70 × 11 = 216.70″

$= 18'\text{–}0\frac{11}{16}''$

After the line length has been determined, it is marked on the top edge of the rafter stock in the same manner as for other kinds of rafters (see Fig. 3-9).

Ridge Length

Theoretical ridge length can be determined by adding the minor run to the length of the building wing (see Fig. 4-5). To determine the actual length of the ridge board, the ridge board must be shortened by an amount equal to one-half the diagonal thickness of the valley rafter plus one-half the thickness of the ridge board (see Fig. 4-6).

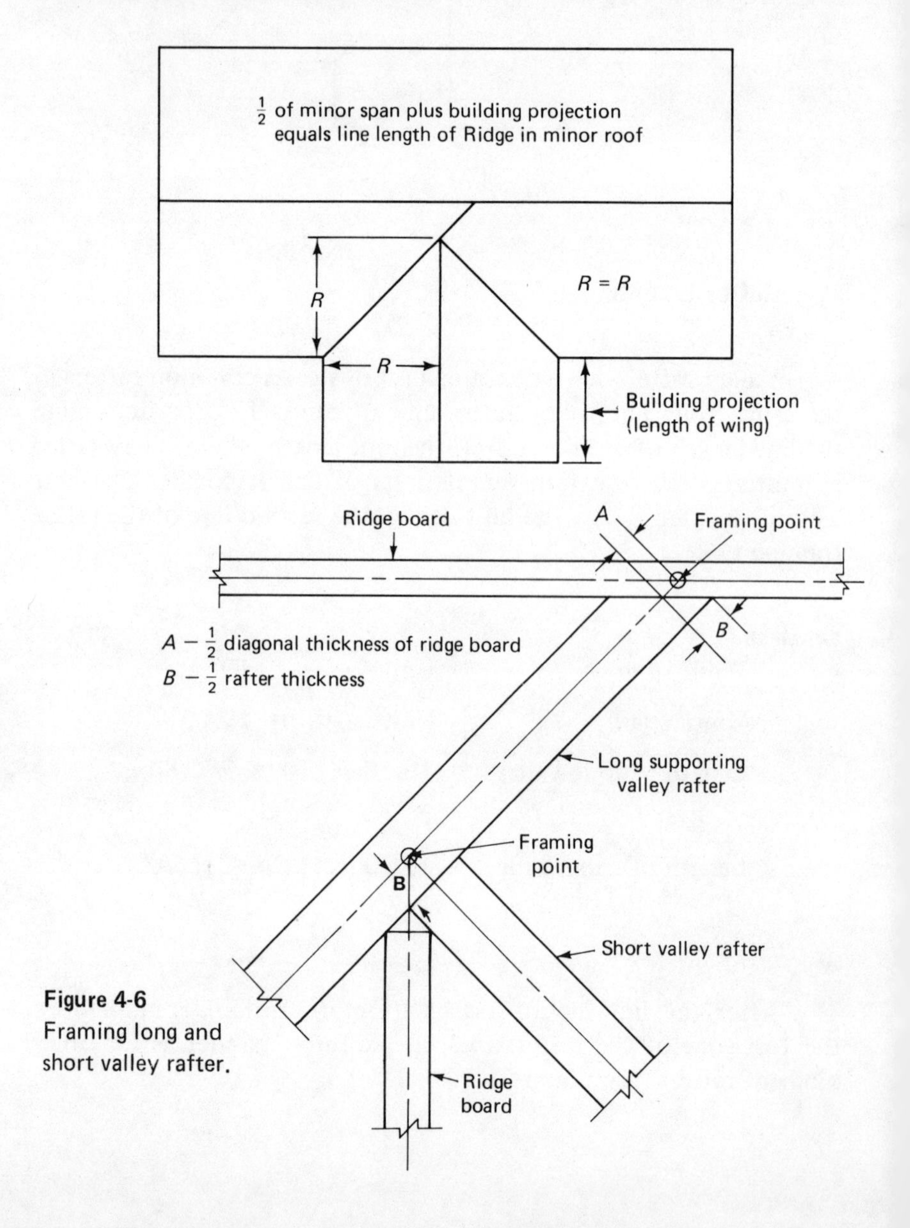

Figure 4-6
Framing long and short valley rafter.

Rafter Layout

The seat cut for long supporting and short valley rafters is laid out in the same manner as described previously under Valley Rafter Seat Cut. Rafter tails are also laid out in the same manner as for other equal pitch valley rafters.

Shortening and Bevels. The amount of shortening for the supporting valley rafter is one-half the diagonal thickness of the ridge board. It is deducted by measuring at right angles to the line-length plumb line. To locate the plumb line at the long point of the bevel, an amount equal to one-half the rafter thickness is measured forward at right angles to the shortening plumb line. The plumb line drawn at the long point of the rafter is the cutting line (see Fig. 4-7). Because the short valley rafter meets the long valley

Figure 4-7
Layout of long valley rafter.

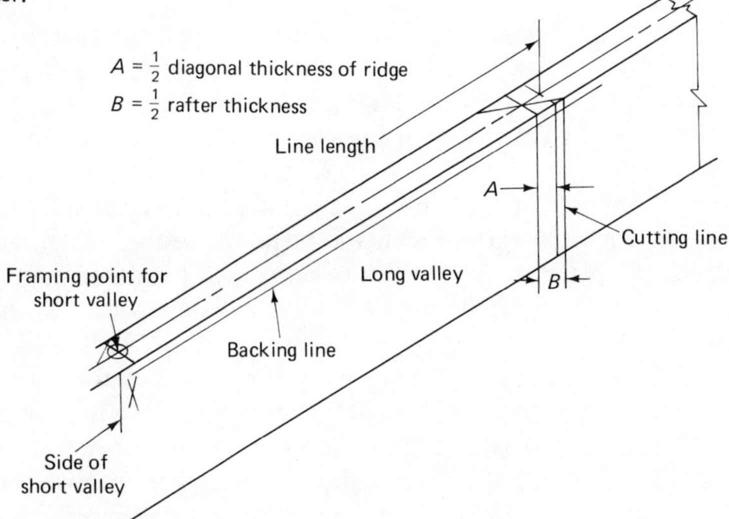

rafter at right angles in plan view, the amount of shortening required is one-half the thickness of the long valley rafter, and the plumb cut does not require a bevel (see Fig. 4-8).

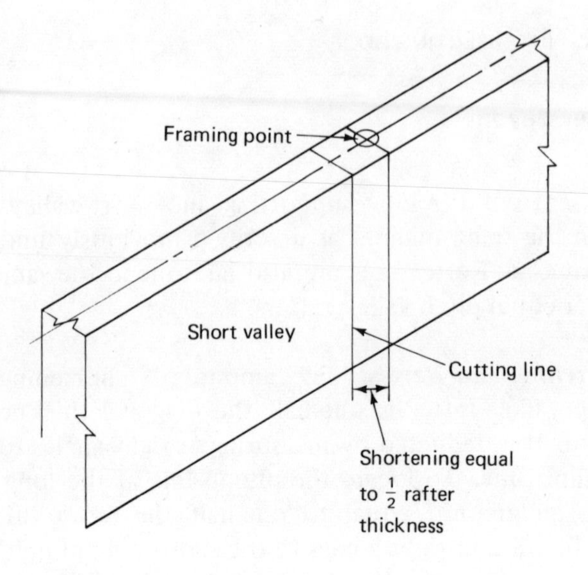

Figure 4-8
Short valley layout.

Backing the Supporting Valley Rafter. The supporting valley rafter must be backed off from the intersection of the short valley up to the ridge board on the long point edge. This backing is required to avoid a high spot in the roof above the intersection of the two rafters. It is needed because the height of the common rafter is established at the center of the rafter and not at the edges. Therefore, when the rafter meets the ridge, the center of the rafter is at ridge height, and the long point edge will be above the ridge unless it is backed off down to the intersection with the short valley.

To determine the amount of backing needed, the framing square is placed on the side of the rafter stock with the unit rise and 17″ held at the edge of the rafter as illustrated in Fig. 4-9. The backing distance is determined by measuring in from the edge of the rafter along the run side of the square a distance of one-half the thickness of the rafter and placing a mark at that point. A line drawn through this point parallel to the edge of the rafter is used as a guide in cutting. The saw is set to follow the line on the side of the rafter and to cut through at the center of the edge of the stock.

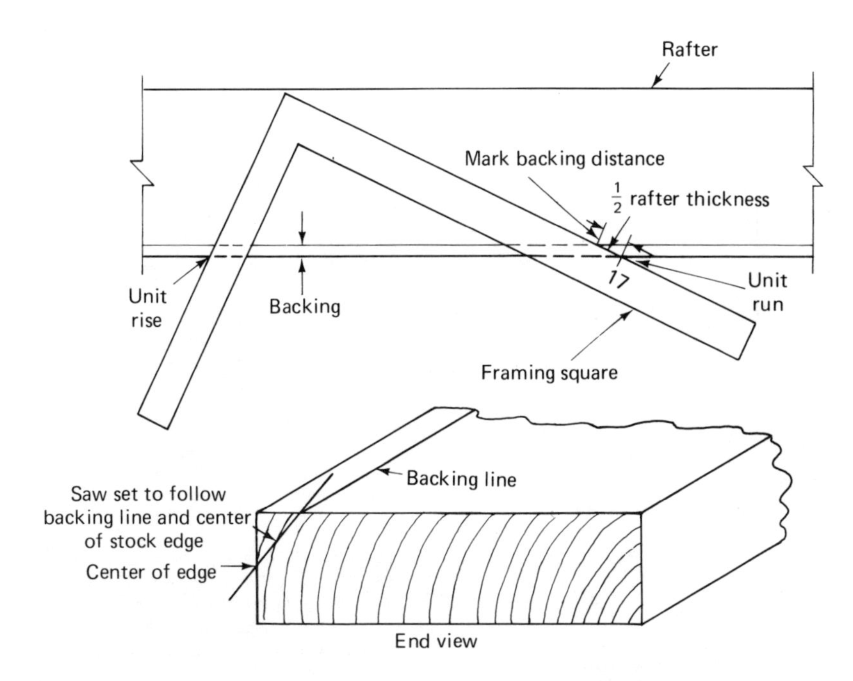

Figure 4-9
Backing a rafter.

VALLEY JACK RAFTERS

The procedure for the layout of valley jack rafters is similar to that for hip jack rafters, the only difference being that the valley jacks do not have a tail, and therefore, the lower end is cut at a bevel to fit against the valley rafter.

To determine the length of the longest valley jack, it is first necessary to find the distance from the far side of the last common rafter to the side of the valley rafter at the building line. By subtracting this distance from the rafter spacing, the amount of run lost by the first jack rafter is determined (see Fig. 4-10).

The long point of the longest valley jack rafter is determined by deducting distance *D* at right angles to the line-length plumb line at the seat cut of the common rafter pattern and drawing a new plumb line at that point. The length of the jack rafter is squared across the top of the common rafter pattern, and the difference in length of jack rafters for the corresponding unit rise and spacing is marked off on the top edge to determine the length of the remaining valley jack rafters (see Fig. 4-11).

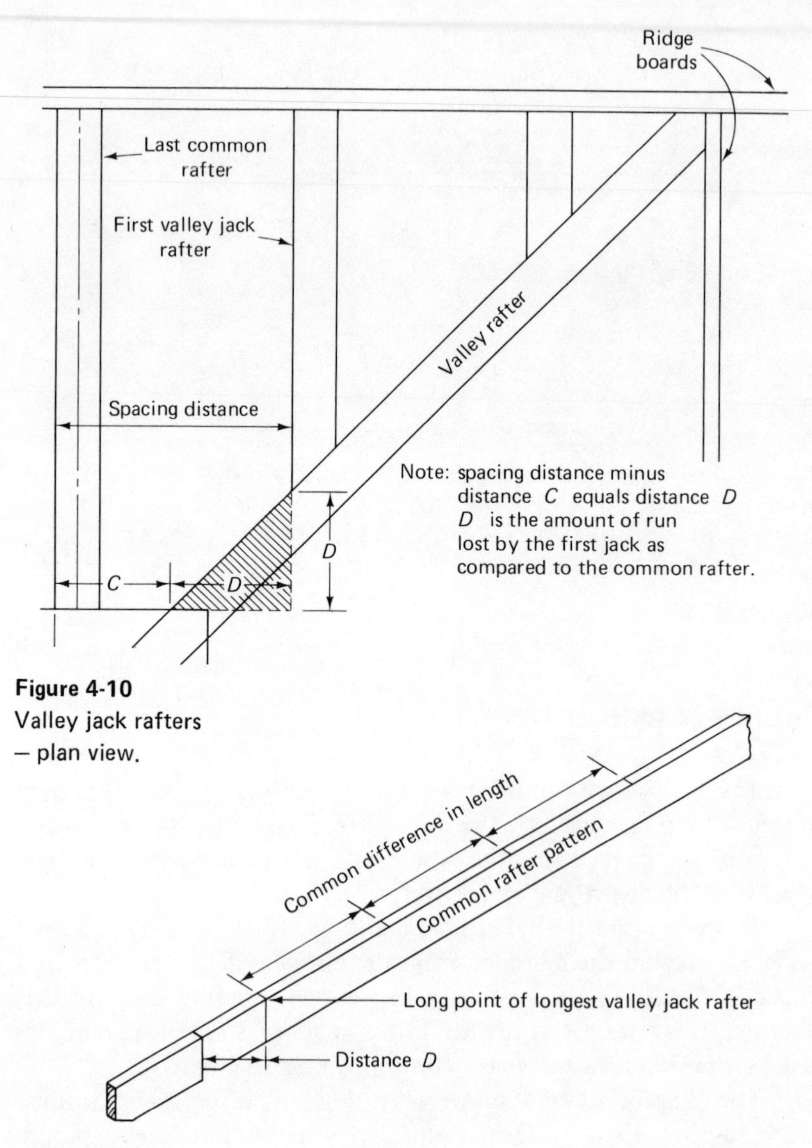

Figure 4-10
Valley jack rafters – plan view.

Figure 4-11
Valley jack layout.

When valley jack rafters are nailed in place, the top edge of the jack rafter is held even with the top of the ridge board, but it is held above the edge of the valley rafter so that the top of the jack rafter is in line with the center of the valley rafter (see Fig. 4-12).

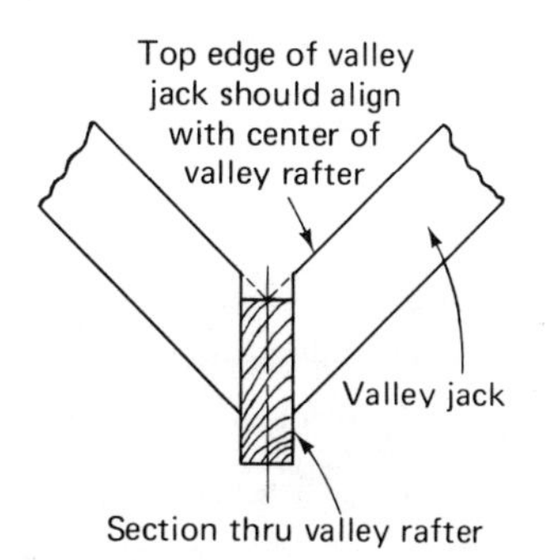

Figure 4-12
Valley jack fastened to valley rafter.

Cripple Jack Rafters

Cripple jack rafters running from a valley rafter to a hip rafter have a run equal to the distance from the center of the hip rafter to the center of the valley rafter measured along the plate line. The line length may be determined by multiplying the common rafter unit length by the number of units of run covered by the cripple rafter.

Line length may also be determined by using the step-off method. When the step-off method is used, one step of common rafter unit rise and unit run is made for each foot of run covered by the cripple rafter. The layout of a cripple jack rafter is illustrated in Fig. 4-13.

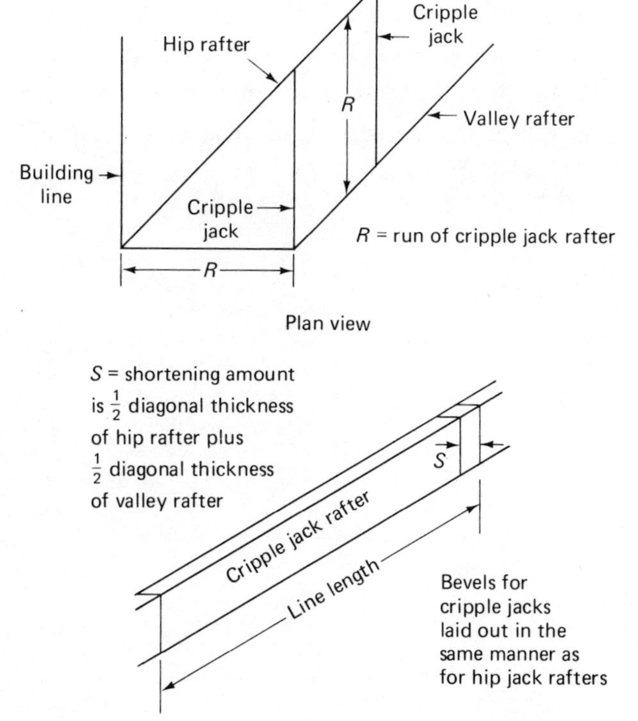

Figure 4-13
Cripple jack rafter layout.

REVIEW QUESTIONS

1. What are the two kinds of intersecting roofs?
2. When is it necessary to have a roof plan?
3. What angle do valley rafters make with the common rafter in plan view?
4. How is the length of valley rafters calculated?
5. Calculate the length of the following valley rafters:

Unit Rise	*Run*
6″	12′
8″	13′
9″	13′–6″
10″	13′–8″
12″	14′–5″

6. Sketch the plan view and rafter layout for equal span valley rafters at the ridge.
7. Sketch the plan view and rafter layout for valley rafter seat cuts.
8. How is the run of a supporting valley rafter determined?
9. How is the run of a short valley rafter determined?
10. How is the length of the ridge board for intersecting roofs determined?
11. Why is it necessary to back long supporting valley rafters?
12. How is the run of valley jacks determined?
13. How is the length of cripple jacks determined?
14. Why is the top edge of valley jacks held above the top of the valley rafter?

FIVE

Unequal Pitch Intersecting Roofs

Intersecting roofs that have the same total rise but different spans are called *unequal pitch roofs.* In the unequal pitch intersecting roof, the valley rafters do not run across the building at an angle of 45° in plan view. The angle at which the unequal pitch valley rafter crosses the building is governed by the run of the two unequal spans and varies as the spans change.

To aid in identifying the various parts of an unequal pitch roof, the span of the main part of the building is called the *major span,* and the span of the intersecting roof is called the *minor span.*

Unequal pitch intersecting roofs of the usual type may be divided into two categories. The first category is the roof which is placed on a building that has no rafter tail projection, and the second is the type which has a cornice projection. The second kind, requiring a rafter tail, involves additional layout work. Therefore, the unequal pitch roof without a cornice, which is easier to lay out and build, will be discussed first.

UNEQUAL PITCH INTERSECTING ROOFS WITHOUT CORNICE

The plan view for an unequal pitch roof without a cornice is illustrated in Fig. 5-1. Line length for the common rafter in the major span is determined by applying the unit method or the step-off method in the usual manner.

The major span is 24′. Therefore, the run is 12′, and with a unit rise of 9″, the line length of the common rafter of the major span is 15′.

Total rise of the main roof is determined by multiplying the unit rise by the number of units of run covered by the main roof. The total rise of the main roof in Fig. 5-1 is 9′–0″ and is also the total rise of the minor roof.

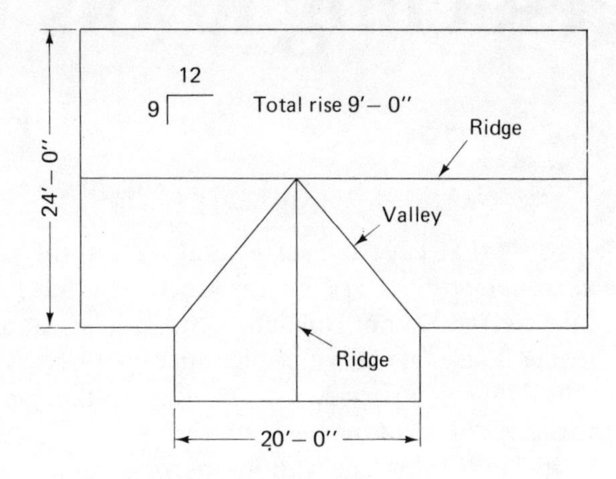

Figure 5-1
Unequal pitch intersecting roof without cornice – plan view.

Common Rafter in Minor Roof

The length of the common rafter in the minor roof is determined by finding the diagonal distance of the total rise and total run of the minor roof. This distance may be scaled on the rafter framing square by using the 1/12 scale on the back of the square. This is done by allowing 1″ for each foot of the rise on one side of the square and for each foot of run on the other side of the square and measuring the diagonal distance.

The step-off method may also be used to determine the line length of the rafter with the 1/12 scale by using the total rise on the tongue of the square and the total run on the body of the square and making 12 steps of rise and run to get the line length. The plumb line is established by marking along the rise side of the square. Level lines are marked along the run side of the square.

Line length may also be determined by applying the Pythagorean theorem.

EXAMPLE:

$$\text{Rafter length} = \sqrt{(\text{total rise})^2 + (\text{total run})^2}$$

$$\text{Rafter length} = \sqrt{(9)^2 + (10)^2} = \sqrt{181}$$

$$\text{Rafter length} = 13'\text{–}5\tfrac{7}{16}''$$

When laying out the common rafter of the minor roof, the line length is marked on the top edge of the rafter in the usual manner. The rafter is shortened by deducting one-half the ridge thickness at right angles to the plumb line, and the height of the common rafter seat cut is made the same as the height of the seat cut for the common rafters of the main roof (see Fig. 5-2).

Figure 5-2
Common rafter for minor roof.

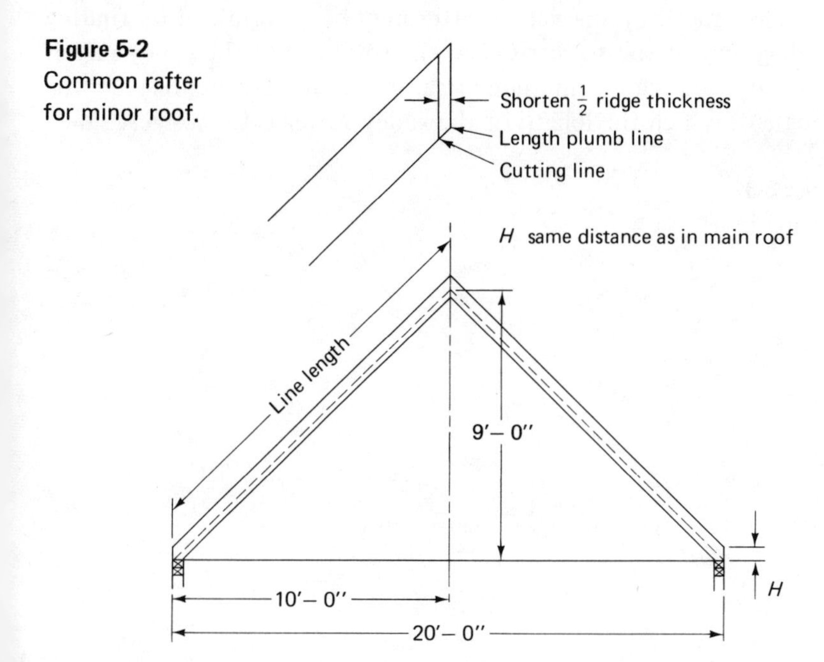

Run of Valley Rafter

The run of the valley rafter for this roof is the diagonal of the major run and the minor run. In most cases, the actual run of the valley rafter may be scaled by using the 1/12 scale on the framing square, but it can also be found mathematically when greater accuracy is necessary.

The valley rafter run for the roof in Fig. 5-1 is determined in the following example.

EXAMPLE:

$$\text{Valley rafter run} = \sqrt{(\text{major run})^2 + (\text{minor run})^2}$$

$$\text{Valley rafter run} = \sqrt{(12)^2 + (10)^2} = \sqrt{244}$$

$$\text{Valley rafter run} = 15'\text{–}7\tfrac{7}{16}''$$

Length of Valley Rafter

The length of the valley rafter may be determined by finding the diagonal of the total rise and the total run of the valley rafter, either by scaling or mathematically. Figure 5-3 represents the triangle of which the length of the valley rafter is the longest side.

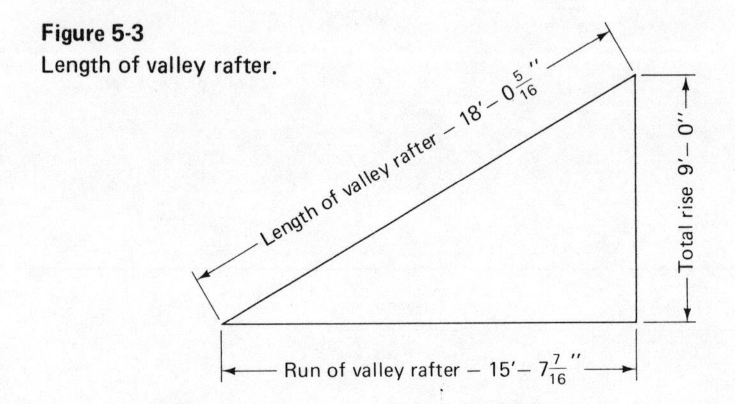

Figure 5-3
Length of valley rafter.

EXAMPLE:

$$\text{Length of valley rafter} = \sqrt{(\text{total rise})^2 + (\text{valley rafter run})^2}$$

$$\text{Length of valley rafter} = \sqrt{(9)^2 + (15'{-}7\tfrac{7}{16}'')^2}$$

$$\text{Length of valley rafter} = \sqrt{81 + 244} = \sqrt{325}$$

$$\text{Length of valley rafter} = 18'{-}0\tfrac{5}{16}''$$

As for any other rafter, the line length of the valley rafter is marked on the top edge of the rafter stock and plumb lines are drawn on the side of the rafter at the line-length mark. Plumb lines for the valley rafter are laid out using the 1/12 scale by holding total rise on the tongue of the square and total valley rafter run on the body of the square. If either the total rise or the total run is too big to fit on the square at 1/12 scale, each may be divided by 2 to get numbers that will fit on the square.

Valley Rafter Bevels

After the line length is marked on the valley rafter, the bevels must be laid out. To determine the shape of the bevels and the amount of shortening required, a plan view of the two runs is drawn to a scale of 1″ equaling 1′ (see Fig. 5-4). The valley rafter run is represented by a broken line as are the center lines of the ridge boards. The actual thicknesses of the ridge boards are drawn in with one-half the thickness on each side of the center line. The valley rafter is then drawn in with one-half its actual thickness on each side of the center line representing its run.

A line that runs through the framing point is squared across the top of "rafter" in plan view, and the distances from the length mark to the center of the ridge board are measured and transferred to the rafter stock. These are all plan-view dimensions and are transferred by measuring at right angles to the plumb lines on the rafter stock (see Fig. 5-4).

The shape of the bevel is drawn on the top edge of the rafter by connecting the framing point with the second plumb line, which has established the depth of the bevel on each side of the rafter.

Figure 5-4
Determining bevels for unequal pitch valley rafters.

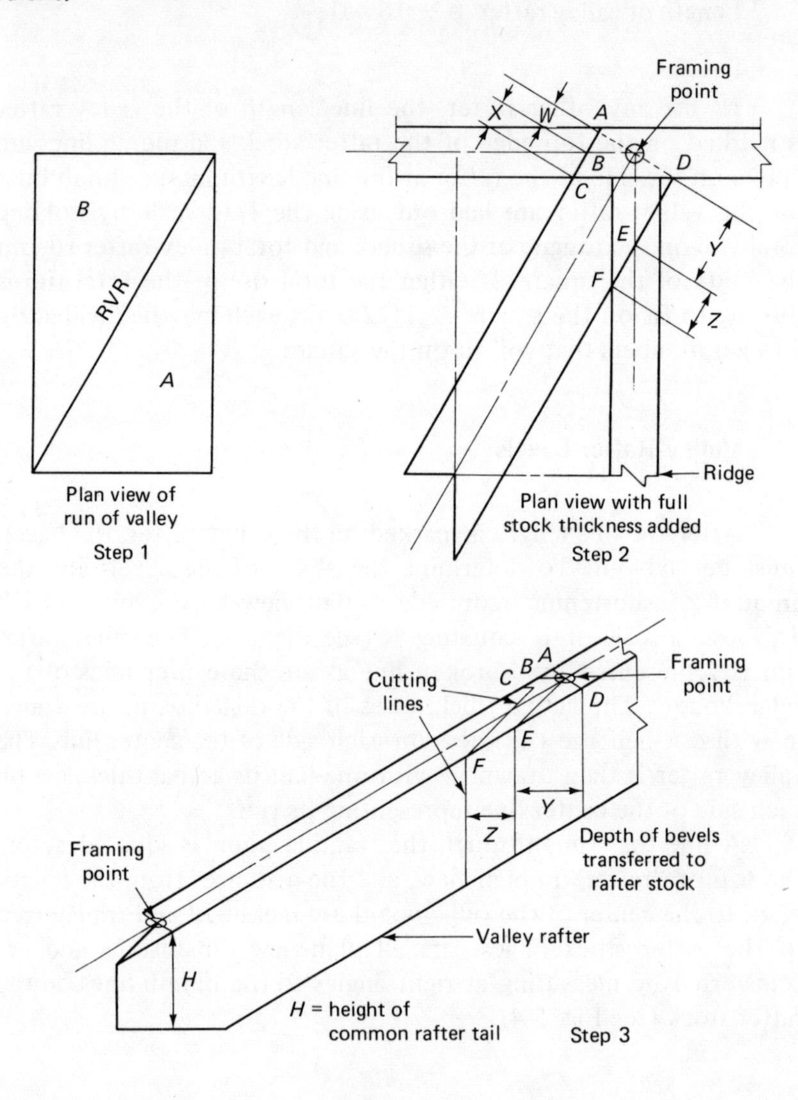

Shortening the Valley. Shortening is accomplished by measuring along the edge of the rafter in plan view, from the center of the ridge to the edge of the ridge. These measurements are transferred to the rafter stock as shown in Fig. 5-4, and new plumb lines are marked on the sides of the rafter. Starting at the final plumb lines, the bevel cutting lines are established by drawing bevels parallel to those originally laid out on the top of the rafter. The long point of the bevel at the cutting lines will not be at the center of the rafter but will be slightly off to one side. The amount that the long point moves off center varies with the various spans of the unequal pitch intersecting roof.

Jack Rafters in Unequal Pitch Roofs

Jack rafters for each section of an unequal pitch roof have a different plumb cut, a different bevel cut, and an uncommon difference in length of jack rafters. The plumb cut is governed by the plumb cut of the common rafters in the section of the roof in which the jack rafters are installed. Jack rafters in the major span have the same plumb cut as common rafters in the major span, and jack rafters in the minor span have the same plumb cut as common rafters in the minor span.

The bevel cut and the difference in length of jack rafters can best be determined by working with the slant triangle. The *slant triangle* is a portion of the roof which is bounded by the valley rafter, a portion of the ridge board, and the common rafter that runs from the valley rafter to the ridge board (see Fig. 5-5).

Figure 5-5
Slant triangles.

LCR – length common rafter
RVR –run valley rafter

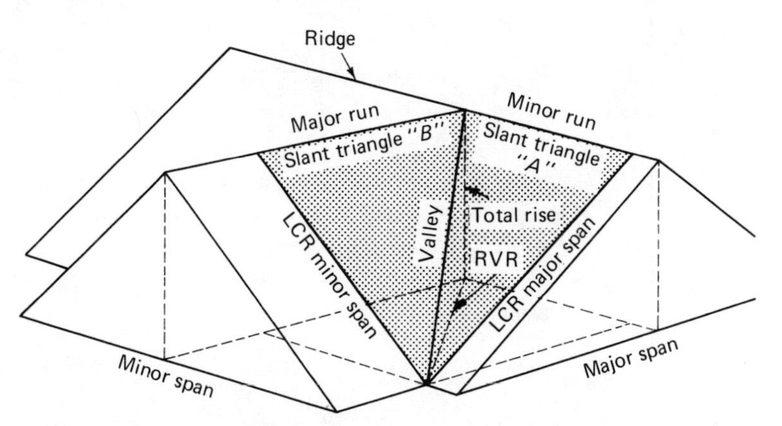

Working with slant triangle *A*, which lies on the main roof, the first step in determining the difference in the length of jack rafters is to find the number of spaces between the rafters included in the ridge length. The length of the ridge in question is that portion cut off by the minor span (see Fig. 5-6). With jack rafters 16″ on center, the number of spaces in the ridge would be 120″ divided by 16″, or $7\frac{1}{2}$ spaces.

To determine the difference in the length of jack rafters, the number of spaces in the ridge is divided into the length of the common rafter in the span under consideration. In Fig. 5-6 the common rafter is 15′–0″ or 180″ long. The difference in length of jack rafters, therefore, is 180″ divided by 7.5 spaces or 24″. Notice how the lines projected from the intersection of the jack and valley rafters divide the common rafter into $7\frac{1}{2}$ parts. In effect, this drawing proves that this method for determining the difference in length of jack rafters is logical.

The same procedure is applied to slant triangle *B* to determine the difference in length of jack rafters for that section of

Figure 5-6
Finding the difference in length of jack rafters.

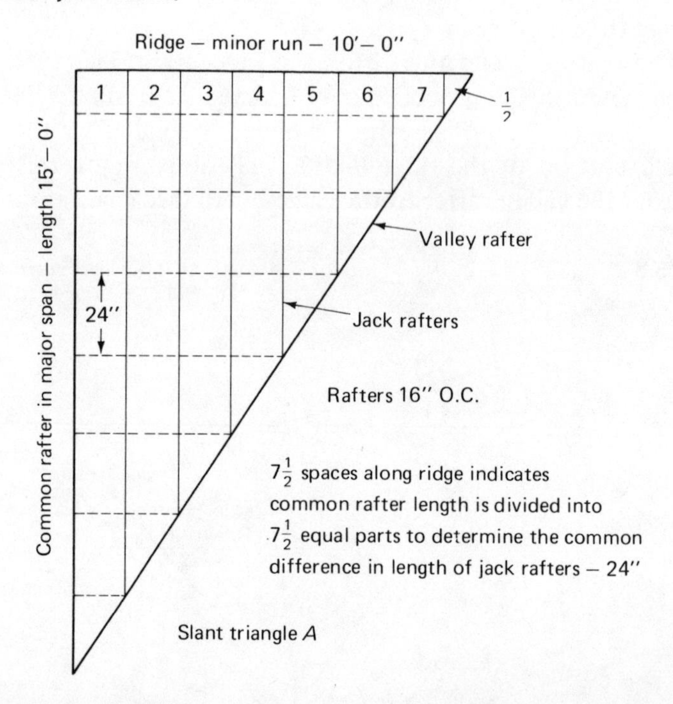

roof. In this triangle the length of the ridge is equal to the major run and the length of the common rafter is taken from the minor span. With the rafters 16″ on center, the ridge length, 144″, divided by 16″ indicates that there are nine spaces contained in the ridge.

The difference in the length of jack rafters is found by dividing the length of the common rafter in the minor span by the number of spaces in the ridge length equal to the major run. In triangle *B* the common difference in jack rafter length is $161\frac{7}{16}''$ divided by nine spaces or $17\frac{15}{16}''$. Notice how the lines projected from the intersection of the valley rafter and jack rafters in Fig. 5-7 divide the common rafter of the minor span into nine equal parts, each $17\frac{15}{16}''$ long.

Jack Rafter Bevels. Bevels on jack rafters may be laid out by using the slant triangles and the 1/12 scale. The bevel on the jack rafter is governed by the length of the common rafter and the length of ridge into which it frames. For jack rafters in the major span, the bevel is controlled by the length of the common rafter in the major span and the length of ridge cut off by the minor run. Therefore, the bevel of jack rafters in triangle *A* in Fig. 5-6 is governed by 15 and 10.

Figure 5-7
Finding the difference in length of jack rafters.

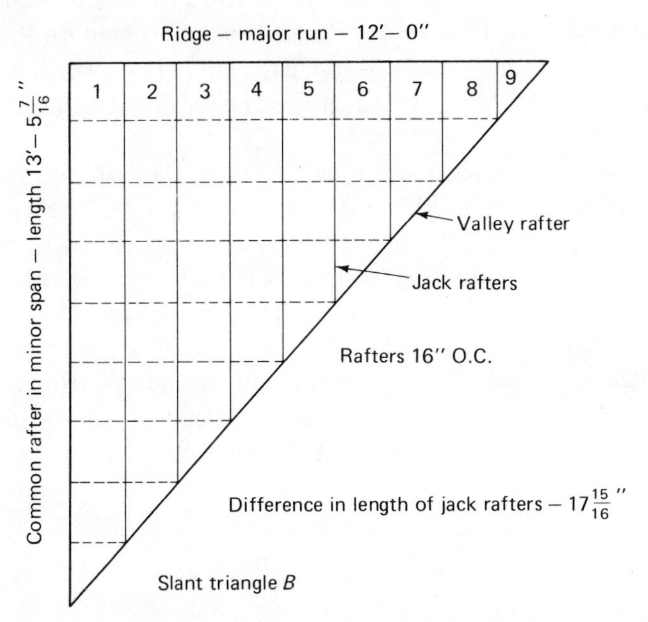

To lay out this bevel on the top edge of the jack rafter, 10″ is held on the tongue of the square and 15″ is held on the body of the square. The bevel is always marked on the side of the square that represents the rafter length, 15″ in this example (see Fig. 5-8).

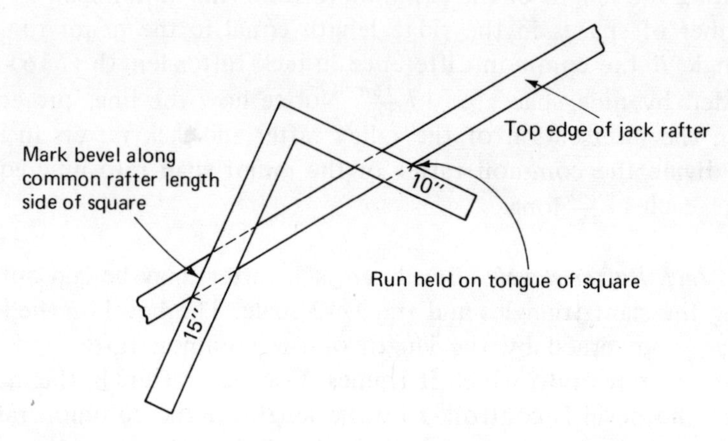

Figure 5.8
Layout of bevels on jack rafters.

The same procedure is applied for the jack rafters in triangle *B* except that the length of the ridge is 12′ and the length of the common rafter is 13′–5$\frac{7}{16}$″. Therefore, 12″ is held on the tongue of the square, and 13 and 5$\frac{1}{2}$ twelfths are held on the body of the square. The bevel is marked along the rafter length side of the square.

Shortening is required when jack rafter lengths are calculated to the center lines of the ridge and valley rafter. The amount of shortening needed at the upper end is the same as that required for a common rafter and is laid out in the same manner. The amount of shortening at the lower end can be found quickly by measuring at right angles to the bevel on the top edge of the rafter stock and marking off a distance equal to one-half the valley rafter thickness. A new bevel is drawn at this point, and a new plumb line is also marked on the side of the rafter (see Fig. 5-9).

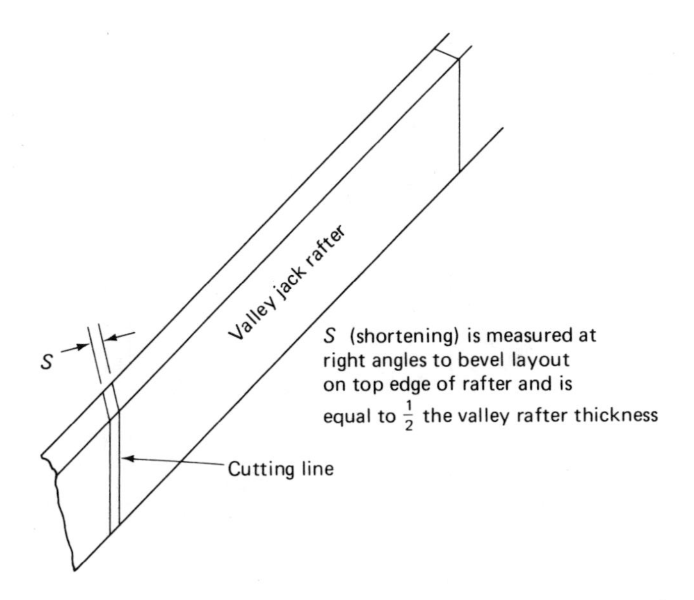

Figure 5-9
Shortening of jack rafters.

UNEQUAL PITCH INTERSECTING ROOF WITH CORNICE

The task of laying out and framing an unequal pitch intersecting roof with a cornice introduces problems not encountered in the roof without a cornice. Generally, there is a desire to maintain an equal width and height in the cornice on both sections of the building, but because one roof is steeper than the other, it will drop farther below the top of the wall than the rafter tail on the flatter roof. To avoid unequal height in the cornice, it is necessary to work from the cornice line when determining the lengths of rafters in the minor roof and to raise the top of the wall for the roof in the minor span.

Figure 5-10 gives the plan view for an unequal pitch intersecting roof with a cornice projection. The length of the common rafter in the main roof and the length of the rafter tail are determined in the usual manner. Before the length of the common rafter in the minor roof can be determined, the cornice drop and total increased rise must be known.

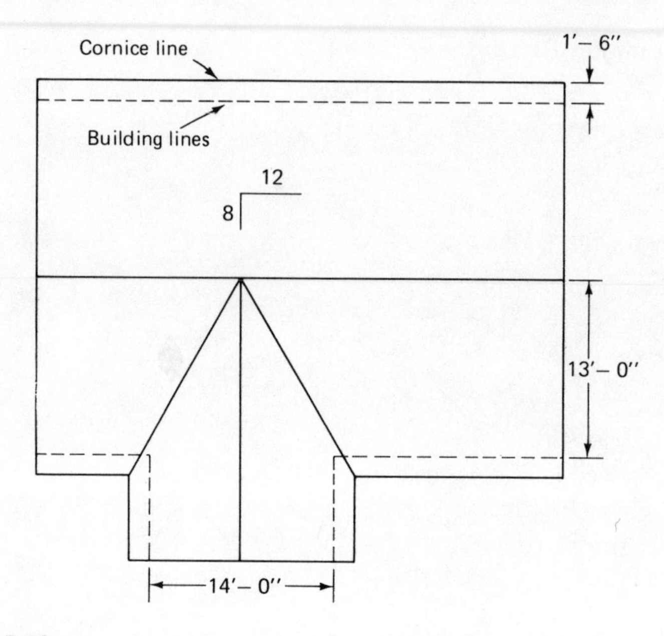

Figure 5-10
Unequal pitch intersecting roof with cornice – plan view.

The cornice drop is determined by multiplying the unit rise of the main roof by the run of the cornice. For the roof in Fig. 5-10, the cornice drop is 1.5 times 8″, or 12″ (see Fig. 5-11).

To determine the total increased rise, the cornice drop is added to the total rise of the main roof. The total increased rise is used to determine the length of the common rafter in the minor roof.

Figure 5-11
Cornice drop.

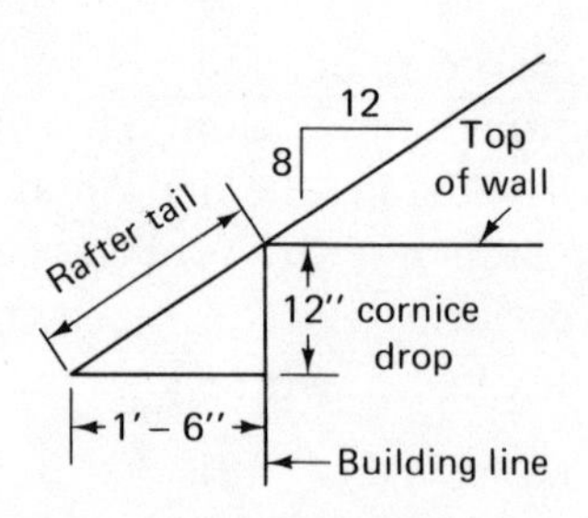

Common Rafter

The total increased run is determined by adding the run of the cornice to the run of the minor span. Total increased run is used in calculating the length of the common rafters in the minor roof.

EXAMPLE:

(See Figs. 5-10 and 5-12)

Total rise of main roof	= 8′–8″
Cornice drop	= 1′
Total increased rise	= 9′–8″
Total increased run of the minor roof	= 8′–6″
Length of common rafter in minor roof	=

$$\sqrt{(8.5)^2 + (9.67)^2} = 12'\text{–}10\tfrac{7}{16}''$$

It should be noted from the comparative drawing in Fig. 5-12 that the common rafter running from the cornice line to the ridge does not meet the top of the normal wall height. Therefore, the plate of the intersecting wall must be raised. The amount that the

Figure 5-12
Minor roof common rafter.

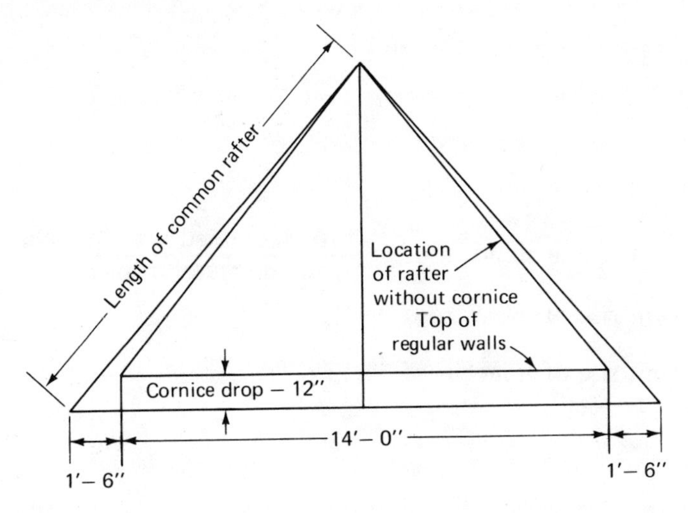

plate must be raised can be determined by making a comparative pitch drawing and measuring the distance between the two rafter slopes (see Fig. 5-13), but it may also be determined mathematically by finding the difference between the cornice rise in the minor and major roofs.

To find the height of the raised plate, the unit rise of the minor roof must be determined. This is done by dividing the total increased rise of the minor roof in inches by the total increased run in feet.

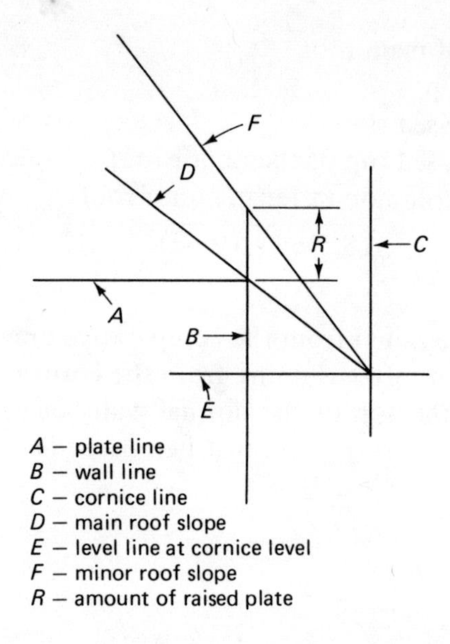

Figure 5-13
Comparative pitch.

EXAMPLE:

$$\text{Unit rise of minor roof} = \frac{\text{total increased rise (in inches)}}{\text{total increased run (in feet)}}$$

$$\text{Unit rise of minor roof} = \frac{116}{8.5}$$

$$\text{Unit rise of minor roof} = 13.65''$$

Rise of minor roof over cornice = 13.65″ × 1.5 = 20.5″

Rise of major roof over cornice = 8″ × 1.5 = 12″

Height of raised plate = 20.5″ - 12″ = 8.5″

The seat cut in the common rafter is located by measuring in from the end of the tail or stepping off the length of the tail from the overall rafter length. When locating the seat cut care must be taken to avoid errors that will cause difficulties in framing the roof. The layout for locating the seat cut is illustrated in Fig. 5-14. Care must be taken to maintain the proper height of the seat cut if the rafter is to fit properly. The height of the plumb cut at the tail must be the same as those on the common rafters of the main roof.

Figure 5-14
Locating common rafter seat cut.

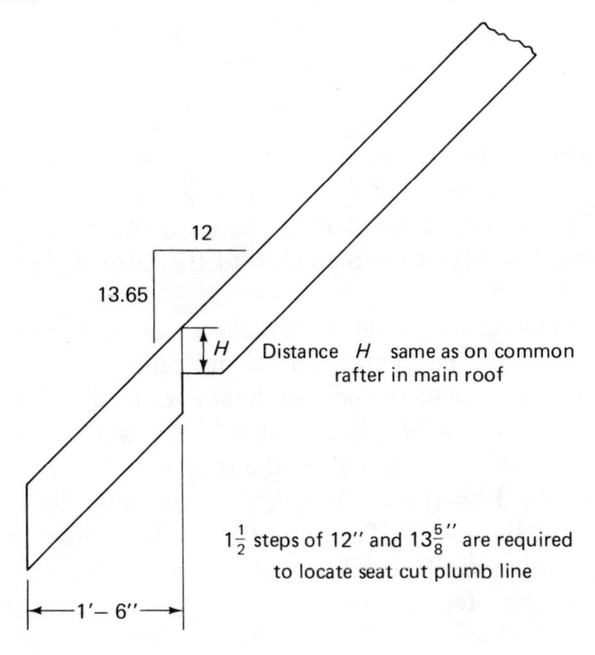

Valley Rafter

The length of the valley rafter is based on the diagonal of the total increased runs of the roof and the total increased rise. To find the run of the valley rafter, the diagonal of the total increased runs may either be scaled or calculated mathematically. The following example is based on the roof in Fig. 5-10.

EXAMPLE:

$$\text{Run of valley rafter} = \sqrt{(\text{increased major run})^2 + (\text{increased minor run})^2}$$

$$\text{Run of valley rafter} = \sqrt{(14.5)^2 + (8.5)^2}$$

$$\text{Run of valley rafter} = \sqrt{282.50} = 16'\text{–}9\tfrac{11}{16}''$$

$$\text{Length of valley rafter} = \sqrt{(\text{increased rise})^2 + (\text{run of valley rafter})^2}$$

$$\text{Length of valley rafter} = \sqrt{(9.67)^2 + 282.50} = 19'\text{–}4\tfrac{11}{16}''$$

Layout of the bevels and shortening for this valley rafter is made in the same manner as for unequal pitch roofs without a cornice. The layout of the seat cut and tail of this rafter requires the use of a drawing of the plan view of the rafter tail made to full scale.

This drawing shows the building lines and cornice lines at the intersecting roof. The center line of the valley rafter is drawn in from the intersection of the cornice lines over to the plate line. The angle at which the center line is drawn is governed by the total increased runs. One run is held on the tongue of the square and the other run is held on the body of the square along the cornice line extended (see Fig. 5-15). The run of the valley rafter tail is measured along the center line in the plan view and is transferred to the valley rafter by stepping off (measuring) at right angles to the plumb line. A plumb line drawn at the distance stepped off represents the length of the rafter tail at its center along the plate line.

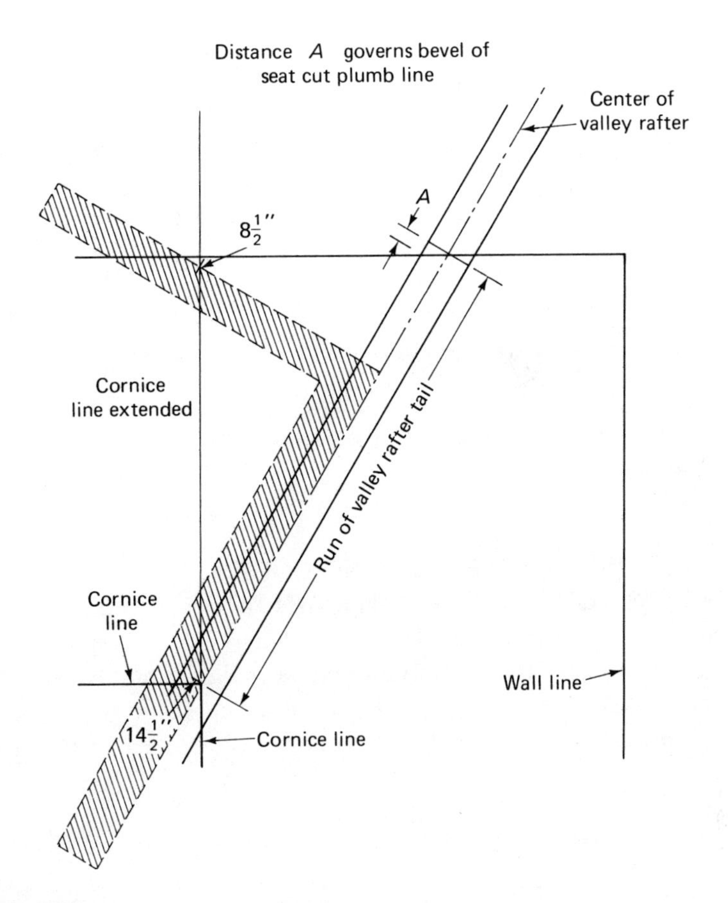

Figure 5-15
Locating valley rafter seat cut.

Because the rafter meets the wall at an angle, the seat cut must be laid out accordingly. The shape of the cut may be taken from the plan view of the valley rafter and transferred to the rafter stock. All plan-view dimensions must be transferred by measuring at right angles to the plumb line (see Fig. 5-16).

Jack rafters for the unequal pitch intersecting roof with a cornice are laid out in the same manner as jack rafters for the unequal pitch intersecting roof without a cornice.

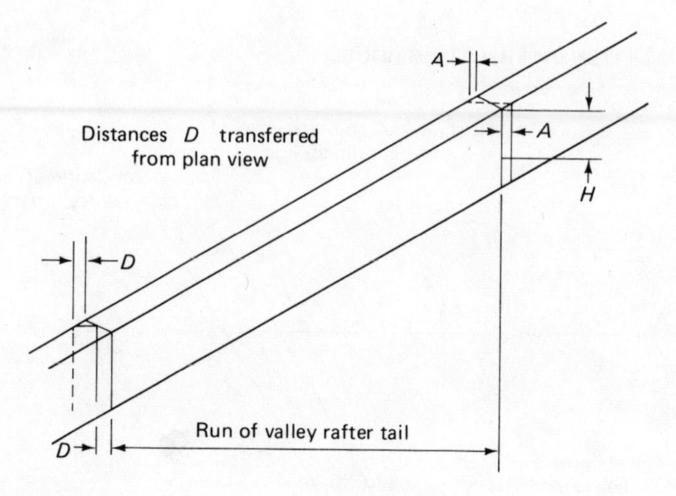

Figure 5-16
Layout of valley
rafter seat cut.

UNEQUAL PITCH INTERSECTING ROOFS WITH LESS TOTAL RISE THAN MAIN ROOF

Occasionally, the carpenter will be called upon to build an unequal pitch intersecting roof that has less total rise than the main roof. The plan view for this roof is illustrated in Fig. 5-17.

Figure 5-17
Wide intersecting roof.

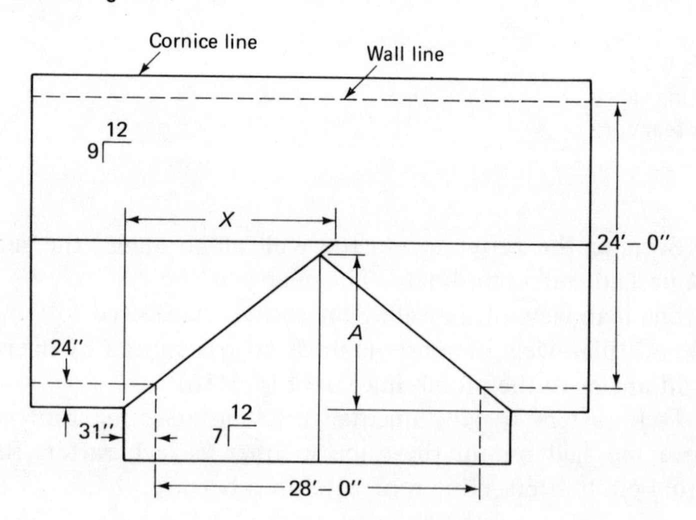

Although unit rise is given for both sections of the roof, there are problems involved in determining cornice width, cornice drop, and plate height. Special problems arise in determining the run, rise, and length of the valley rafters.

Twelve steps that may be used to solve this roof problem in a logical manner are listed and explained in the following:

1. Find cornice drop of the main roof. The cornice drop is determined by multiplying the cornice projection by the unit rise of the main roof.
2. Find cornice drop of the minor roof. The cornice drop of the minor roof is determined by multiplying the cornice projection of the minor roof by the unit rise of the minor roof. Often the cornice width of the minor roof may be adjusted to provide a cornice drop equal or nearly equal to the cornice drop of the main roof.
3. Find the height of the raised or lowered plate. The difference in plate height is determined by finding the difference between the two cornice drops. If the cornice drop of the minor roof is greater than that of the major roof, the plate of the minor roof must be raised, but if it is less than that of the major roof, it must be lowered. If the difference between the plate height is small, the plates may be left at the same height, and the difference in cornice drops may be made up in the height of the seat cuts of the common rafters.
4. Find the length of the common rafter in the major span. This can be done in the usual manner, but to make calculating the lengths of the valley rafters easier, the total length of the rafter (including the tail) should be determined.
5. Find the length of the common rafter in the minor span. Once again, this can be done in the usual manner, but to facilitate calculating valley rafter lengths, the total length of the rafter (including the tail) should be determined.
6. Find the length of ridge *A* in the minor span (see Fig. 5-17). The height of the minor ridge must be determined first. Once the total height of the minor roof, in inches, is known, it is divided by the unit rise of the major roof. The result is the number of units of rise required in the main roof to reach the

height of the minor ridge. There is the same number of units of run included in the distance labeled A. Therefore, to determine distance A, the total rise of the minor roof is divided by the unit rise of the major roof.

7. Find the length of the short valley rafter. The length of the short valley rafter may be determined by using the slant triangle method. The short valley rafter is the hypotenuse of the triangle which has distance A for one leg and the overall length of the minor common rafter for the other leg. The rafter length may either be scaled or determined mathematically.
8. Find the length of the long valley rafter. The sides of similar triangles are proportional. Therefore, the length of the long valley rafter is proportional to the length of the short valley rafter, and distance A is proportional to the distance covered by the total increased run of the major roof. A proportion for determining the length of the long valley rafter could be set up as follows:

$$\frac{\text{Length of long valley rafter (LVR)}}{\text{Length of short valley rafter (SVR)}} = \frac{\text{Major increased run (MaR)}}{A}$$

$$\text{LVR} = \frac{\text{(SVR) (MaR)}}{A}$$

9. Find the run of the short valley rafter. The run of the short valley rafter is the diagonal of distance A and the total increased run of the minor roof. It may be determined either by scaling or mathematically.
10. Find the total rise of the short valley rafter. The rise of the short valley rafter is the same as the total increased rise of the minor roof. It is determined by multiplying the unit rise of the minor roof by the total increased run of the minor roof.
11. Find the cut of the valley rafters. The cut of the valley rafters is determined by the numbers held on the framing square when drawing the plumb lines and level lines. It is governed by the total rise and total run of the valley rafter or a proportional part of the total rise and total run.

12. Find the run X of the long valley. The run X of the long valley is the distance measured along the ridge from the framing point at the cornice intersection to the point where the center line of the valley meets the center line of the ridge. Run X is proportional to the total increased run of the minor roof. Therefore, the following proportion may be set up to determine the X distance:

$$\frac{X}{\text{Increased minor run (IMiR)}} = \frac{\text{Increased major run (IMaR)}}{A}$$

$$X = \frac{\text{(IMiR) (IMaR)}}{A}$$

To illustrate the use of the preceding 12 steps of solving the unequal pitch roof that has less total rise than the main roof, the following example based on the plan view in Fig. 5-17 is given:

EXAMPLE:

1. Find the cornice drop. Main roof: $2 \times 9'' = 18''$.
2. Find the cornice drop. Minor roof: $2\frac{7}{12} \times 7'' = 18\frac{1}{12}''$.
3. Find the height of raised or lowered plate. $18\frac{1}{12}'' - 18'' = \frac{1}{12}''$ raised.

NOTE: Because of the small difference between the two cornice drops, the difference will be made up in the height of the seat cut.

4. Find the length of the common rafter, major span (including tail). $14 \times 15'' = 17'\text{–}6''$.
5. Find the length of the common rafter, minor span (including tail). $16.58 \times 13.89'' = 19'\text{–}2\frac{5}{16}''$.
6. Find the length of ridge A.

$$\frac{16\frac{7}{12} \times 7''}{9} = 12.9 \text{ units or } 12'\text{–}10\frac{13}{16}''$$

7. Find the length of the short valley rafter.

$$\sqrt{(12.9)^2 + (19.19)^2} = 23'\text{–}1\tfrac{1}{2}''$$

8. Find the length of the long valley rafter.

$$\text{LVR} = \frac{(\text{SVR})\,(\text{MaR})}{A} = \frac{(23.12)\,(14)}{12.9} = 25'\text{–}1\tfrac{1}{8}''$$

9. Find the run of the short valley.

$$\text{RSV} = \sqrt{(12.9)^2 + (16.58)^2} = 21'\text{–}0\tfrac{1}{8}''$$

10. Find the rise of the short valley rafter.

$$16\tfrac{7}{12}'' \times 7'' = 116\tfrac{1}{12}'' = 9'\text{–}8\tfrac{1}{12}''$$

11. Find the cut of the valley rafters.
    From 9 and 10 above, rise = $9\frac{8}{12}''$ and run = 21″.
    Plumb lines are marked on $9\frac{8}{12}''$.
    Level lines are marked on 21″.
12. Find run X of the long valley rafter.

$$X = \frac{(\text{IMiR})\,(\text{IMaR})}{A} = \frac{(16.58)\,(14)}{12.9} = 18'\text{–}0''$$

To lay out the bevels for the valley rafter, a plan-view drawing of the runs involved is made to a scale of 1″ equaling 1′. The thicknesses of the ridges and valleys are drawn in full size with one-half the thickness on each side of the center line in the plan-view drawing. This drawing is illustrated in Fig. 5-18. All dimensions transferred from plan view to the rafter stock must be transferred by measuring at right angles to the rafter plumb line.

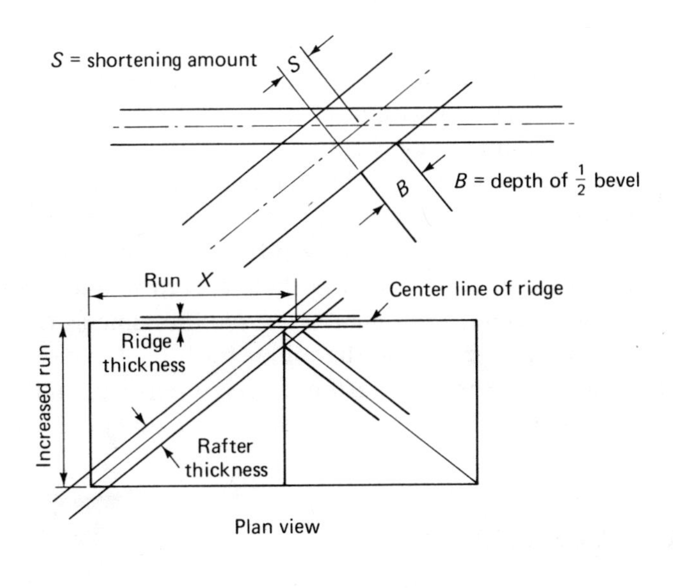

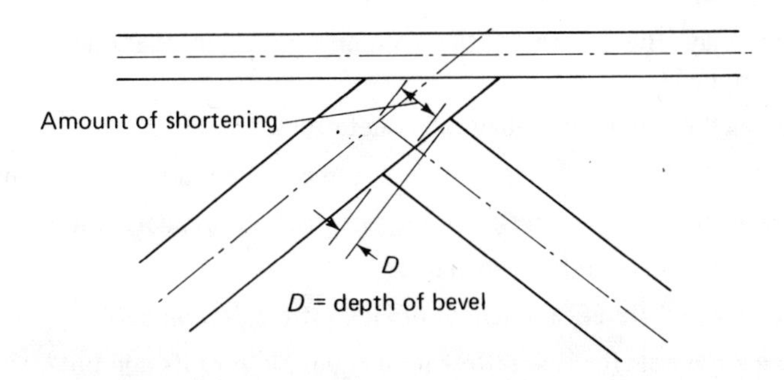

Figure 5-18
Layout of rafters for wide intersecting roof.

The difference in the length of jack rafters in the major span is determined by first dividing distance *X* by the rafter spacing (12″, 16″, or 24″) to determine the number of spaces. The number of spaces is divided into the overall length of the common rafter to get the difference in the length of jack rafters.

To find the difference in the length of jack rafters in the minor span, the number of spaces in distance *A* is divided into the overall length of the minor span common rafter.

The bevel on the jack rafters in the main roof is governed by the length of the common rafter of the main roof and distance *X*. The bevel is marked on the common rafter length side of the square.

In the minor roof the bevel on the jack rafters is governed by the length of the common rafter in the minor roof and distance *A*. The bevel is marked on the common rafter length side of the square.

The plumb lines on jack rafters in the major roof are the same as those for the common rafters in the major roof, and the plumb lines on the jack rafters in the minor roof are the same as the plumb lines on the common rafters in the minor roof.

Shortening of the jack rafters is accomplished in the same manner as for jack rafters in other kinds of unequal pitch roofs.

REVIEW QUESTIONS

1. What is an unequal pitch intersecting roof?
2. How may the length of the common rafter in the minor span be determined?
3. How is the run of the valley rafter determined?
4. How is the length of valley rafters for unequal pitch roofs determined?
5. How is the shape of bevels for unequal pitch rafters determined?
6. How is the valley rafter shortened?
7. How is the difference in length of jack rafters determined?
8. How are bevels for jack rafters in unequal pitch roofs laid out?
9. How does the layout for the unequal pitch roof with a cornice differ from the layout for the roof without a cornice?
10. Outline the procedure for laying out an unequal pitch roof when the intersecting roof has less total rise than the main roof.

SIX

Special Conditions

The number of unusual conditions the carpenter may encounter is without limit. However, if he has a thorough understanding of basic roof framing principles, he can apply them to solve any unusual problem. The following pages present a method for framing octagon roofs, deck roofs, and ornamental double gable roofs. Information is also presented on layout of gable studs, on difference in length of jack rafters, and on bevel layout for jack rafters.

OCTAGON ROOF

The octagon roof is used on small ornamental structures and also to cover an octagon bay on a structure. It consists of hip rafters, common rafters, and jack rafters. Common rafters and jack rafters have the same unit of run in octagon roofs as in other kinds of roofs. However, the hip rafters in an octagon roof run into the building at an angle of $22\frac{1}{2}°$ to the common rafters. Therefore, a new unit of run must be established for the hip rafter. This can be done by laying the framing square on the plan view of the octagon with the body of the square parallel to the common rafter run. Twelve inches on the body of the square is held at the hip rafter center line and a mark is placed where the tongue crosses the center line (see Fig. 6-1).

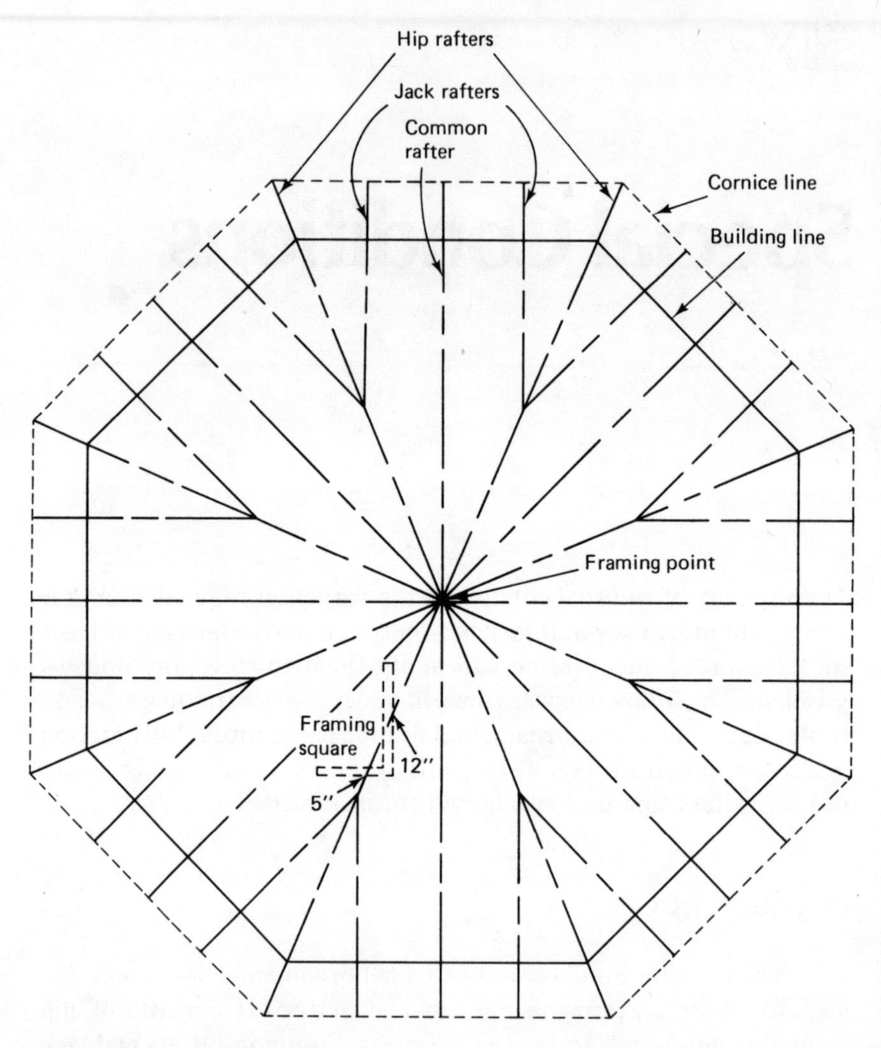

Figure 6-1
Octagon roof — plan view.

The tongue appears to cross the center line at 5″ and the diagonal distance measured along the center line of the hip rafter is 13″. This distance is the unit run of the octagon roof hip rafter and can be used to lay out all octagon roof hip rafters.

If the unit run of the octagon roof were calculated mathematically, it would be necessary to use the trigonometry tangent function. The known angle is $22\frac{1}{2}$ degrees, and the known side is 12″ (see Fig. 6-2). By applying the tangent of $22\frac{1}{2}°$ (.4142) taken from a table of natural trigonometric functions, side a is determined to be 4.97″.

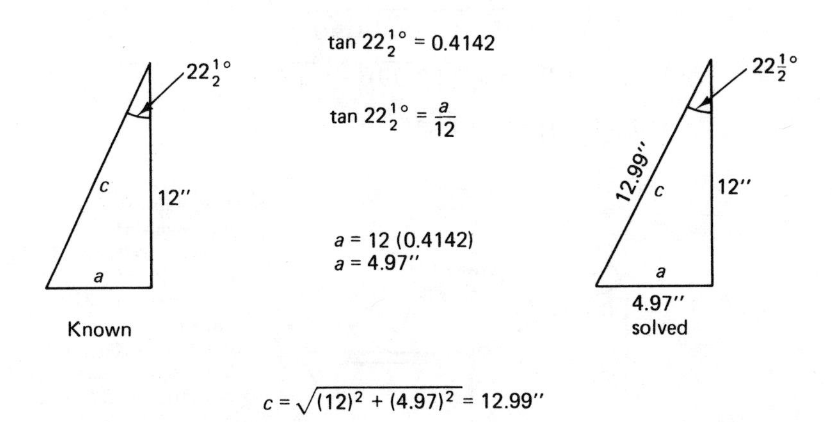

$$c = \sqrt{(12)^2 + (4.97)^2} = 12.99''$$

Figure 6-2
Calculating unit run of octagon roof hip rafter.

The length of side c is easily determined by applying the Pythagorean theorem:

$$c^2 = (12)^2 + (4.97)^2$$
$$c^2 = 144 + 24.70$$
$$c = \sqrt{168.70}$$

(See Chapter 1 for discussion on determining square root.)

$$c = 12.988, \text{ or } 12.99''$$

As the calculations show, for all practical purposes, 13″ can be used as the unit run for octagon roof hip rafters.

The unit rise of the octagon hip rafter is the same as that used for the common rafter (see Fig. 6-3). The unit length of the hip rafter can be calculated by applying the Pythagorean theorem:

$$\text{Unit length of octagon hip rafter} = \sqrt{(\text{unit run})^2 + (\text{unit rise})^2}$$

$$\text{U.L.O.H.R.} = \sqrt{(12.99)^2 + 6^2} = \sqrt{204.70}$$

$$\text{U.L.O.H.R.} = 14.31''$$

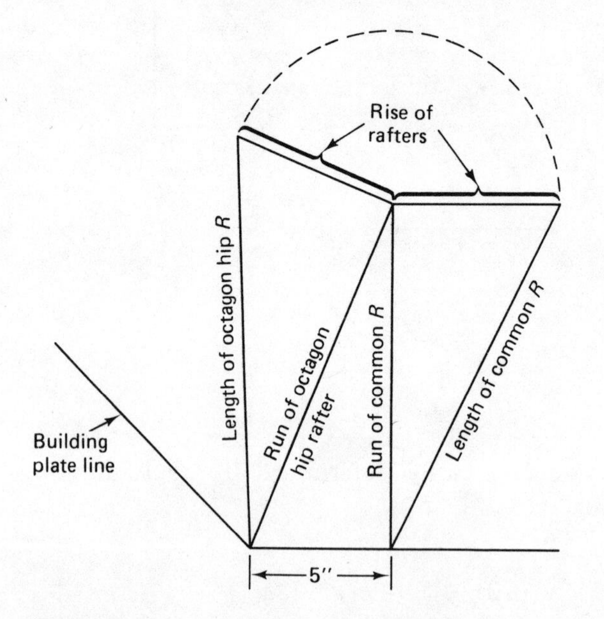

Figure 6-3
Relationship of octagon hip rafter to common rafter.

The unit length of the rafter can be scaled by holding the unit rise on the tongue of the square, the unit run on the body of the square, and measuring the diagonal distance (see Fig. 6-4). In this example the hip rafter with a unit rise of 6″ has a unit length of $14\frac{5}{16}$″. The line length of the rafter is determined by multiplying the unit length by the number of units of run.

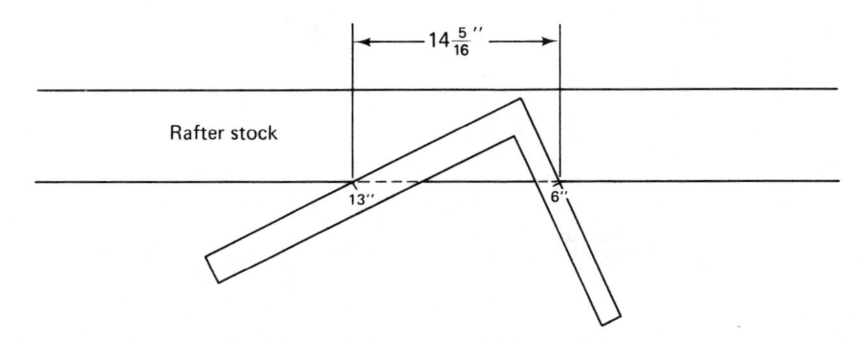

Figure 6-4
Scaling unit length of octagon hip rafter.

Using the roof in Fig. 6-1 as an example, we see that the run of the common rafter is 6′, and the run of the hip rafter is six units. The length of the hip rafter is six units long: 6 × 14.31″ = $85\frac{7}{8}$″. The line length is marked on the rafter stock in the same manner as for other kinds of rafters (see Fig. 6-5). The length lines are squared across the top edge of the rafter, and plumb lines are drawn on the side by holding unit rise (6″) on the tongue of the square, the unit run (13″) on the body of the square, and marking on the unit rise side.

Figure 6-5
Line length marked on stock.

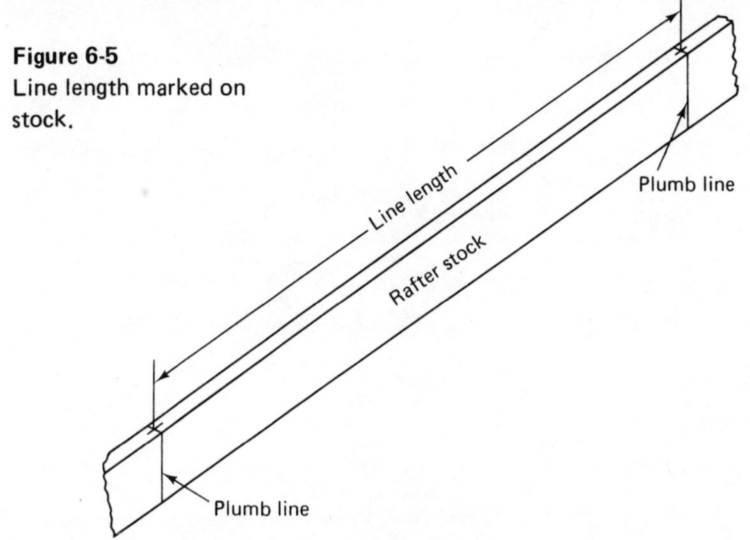

After the line length is marked on the stock, it is necessary to shorten the rafter for the ridge. The easiest way to frame octagon roofs at the peak is to use a 4 X 4 ridge that has been cut into octagon shape (see Fig. 6-6). The amount of shortening required is

Figure 6.6
Framing octagon hip rafters at ridge.

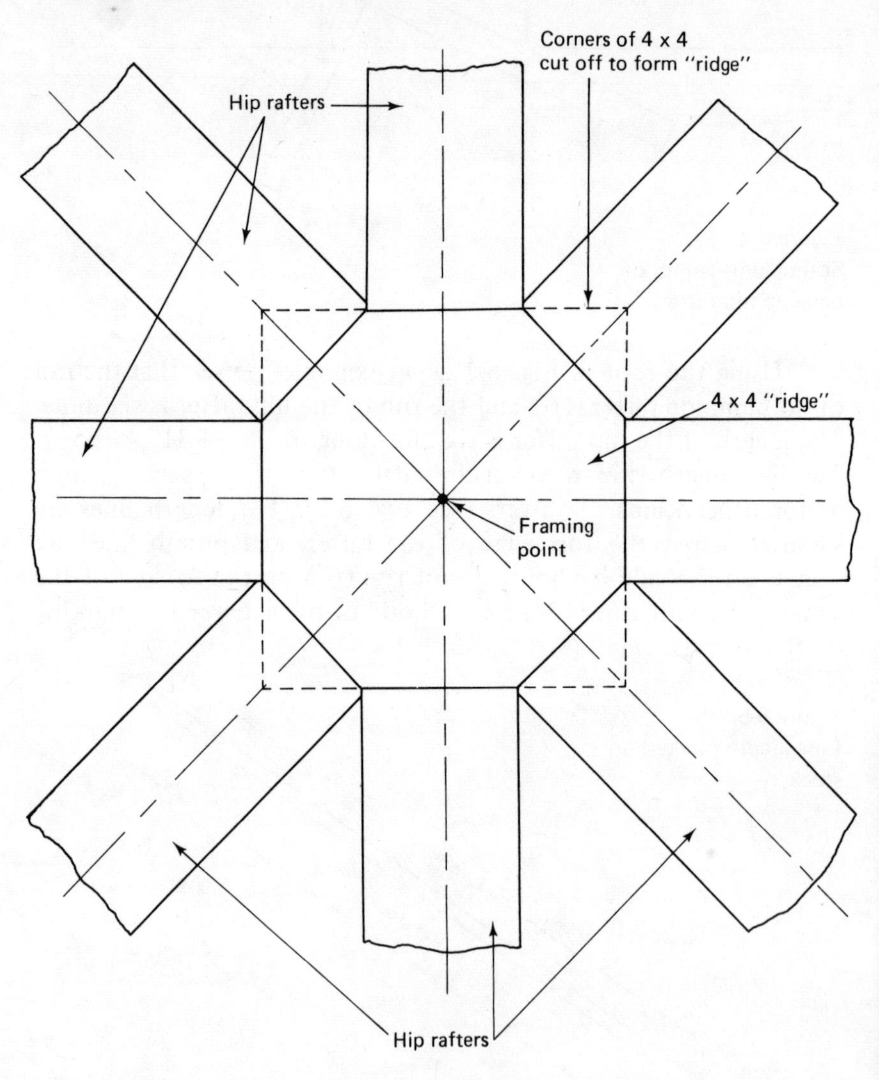

one-half the width of the 4 X 4 ($1\frac{3}{4}''$). Because this is a plan-view dimension, it is transferred to the rafter by measuring at right angles to the plumb line (see Fig. 6-7).

Figure 6-7
Shortening octagon hip rafter.

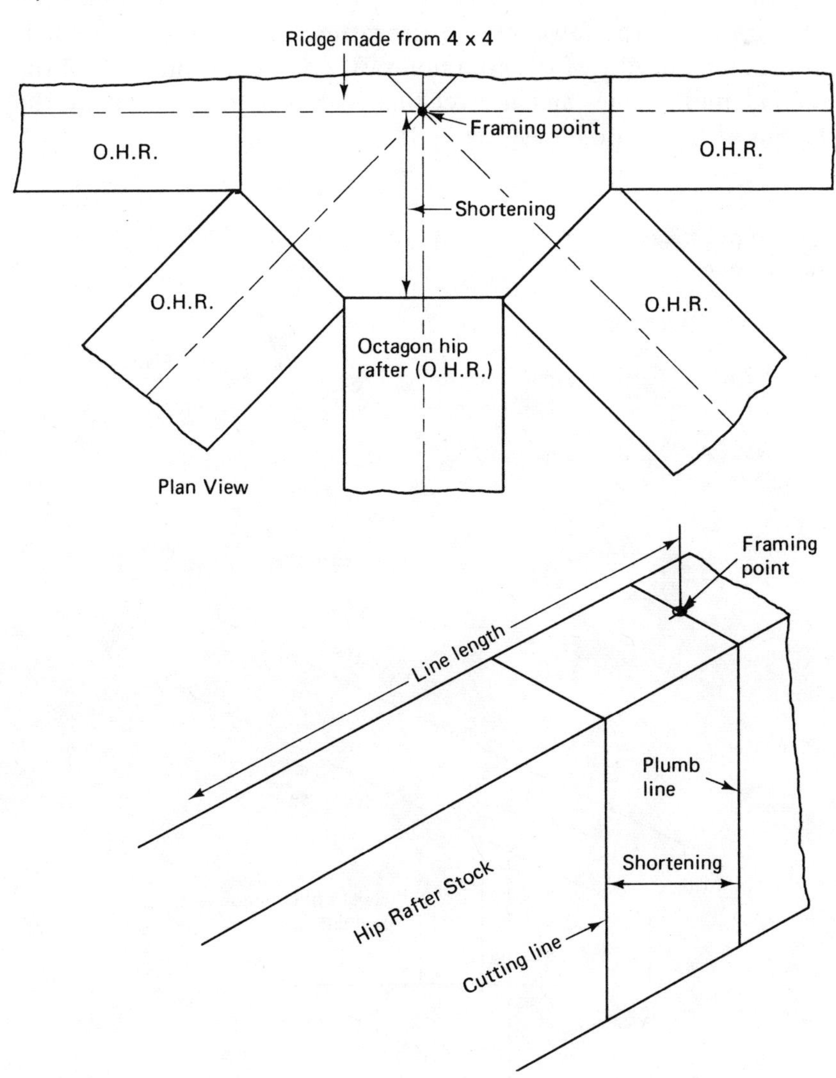

The seat cut of the rafter is laid out in the manner shown in Fig. 6-8. The hip rafter must be set at the same height as the common rafter. Therefore, the point where the edge of the hip rafter crosses the plate line must be determined, and its location must be marked on the side of the hip rafter. The plan view shows that the edge of the rafter crosses the platc $\frac{5}{16}''$ away from the line-length mark. This distance is measured in at right angles to the plumb line on the lower end of the rafter, and a new plumb line is drawn. The height of the common rafter seat cut is marked on the second plumb line, and a level line is drawn in to complete the layout of the seat cut (see Fig. 6-8).

Figure 6-8
Layout of octagon hip rafter seat cut.

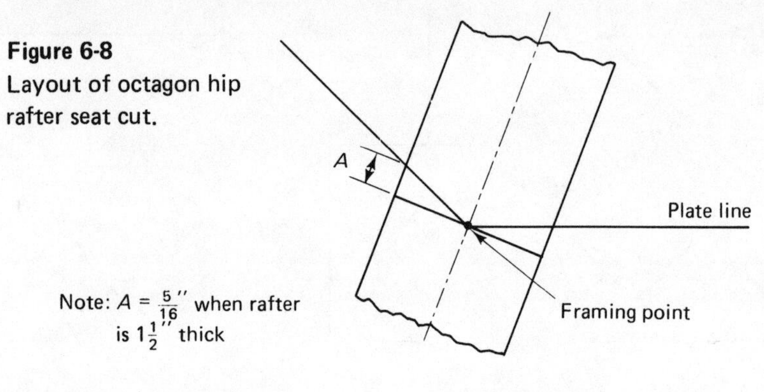

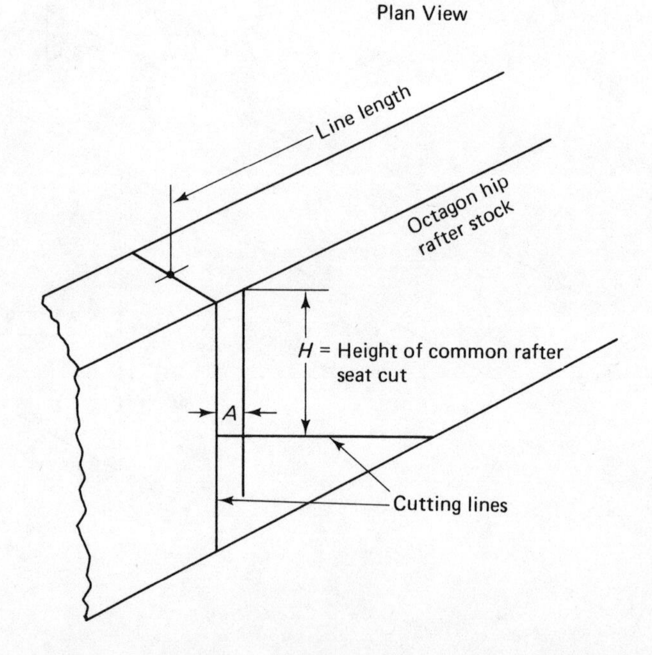

The tail of the rafter is laid out in accordance with the cornice detail. The hip rafter tail has the same number of units of run as the common rafter tail. The roof in Fig. 6-1 has a 12″ projection. Since this is one unit of run, the hip rafter tail has a 13″ run. This plan-view dimension is marked on the side of the rafter as shown in Fig. 6-9.

Figure 6-9
Layout of octagon hip rafter tail.

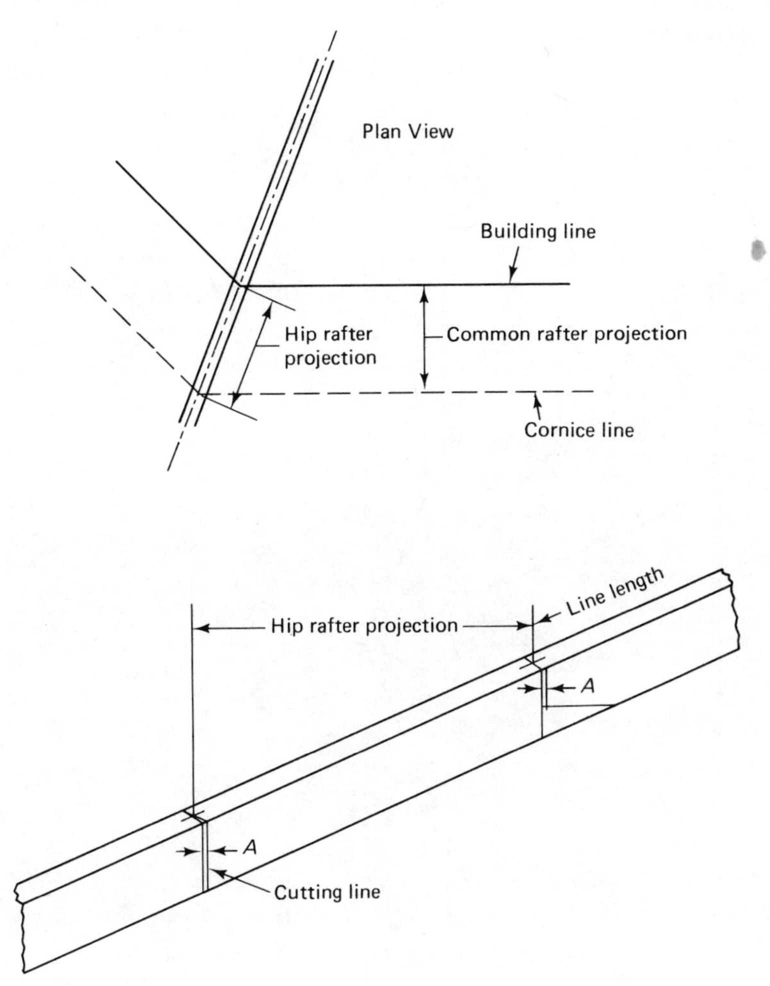

Common Rafters in an Octagon Roof

If the common rafter is to be framed to the ridge, it would be necessary to make a very acute double bevel at the upper end. This would be difficult to accomplish, and it would also be difficult to fasten the rafter in place. Therefore, it is more practical to place a header between the hip rafters 12″ to 24″ down from the peak to which the common rafters can be fastened (see Fig. 6-10).

Figure 6-10
Framing plan.

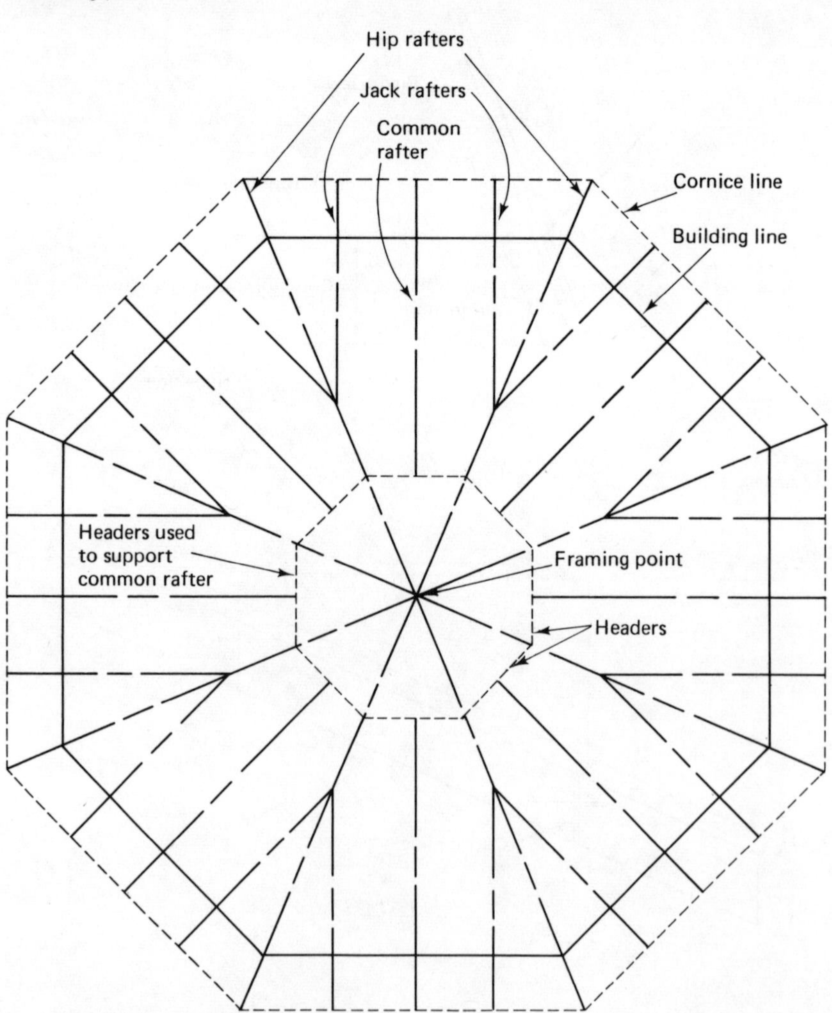

The length of the common rafters in an octagon roof is based on the actual run of the rafter. The run of the rafter is the plan-view distance from the wall to the header. In our example the actual run of the rafter is 4′–0″. For further discussion on determining the length of common rafters, see Chapter 2.

Jack Rafters in an Octagon Roof

Jack rafters in an octagon roof have the same unit run, unit rise, and unit length as the common rafter in the same roof. Because these rafters are usually placed symmetrically around the building, it is generally easy to determine the actual run of the rafter by applying either the trig tangent function for $22\frac{1}{2}°$ (.4142) or the trig cotangent function for $22\frac{1}{2}°$ (2.4142). To use these functions, first the distance from the edge of the hip rafter to the far side of the jack rafter must be measured. In our example that distance is $13\frac{13}{16}''$, or 13.81″.

By applying the tangent function:

$$\text{Run of jack rafter} = \frac{\text{distance from hip to jack}}{.4142}$$

$$\text{Run of jack rafter} = \frac{13.81''}{.4142} = 33.34''$$

By applying the cotangent function:

$$\text{Run of jack rafter} = 2.4142 \times \text{distance from hip to jack}$$
$$\text{Run of jack rafter} = 2.4142 \times 13.81'' = 33.34''$$

Many prefer to use the cotangent function for determining run because it is easier to multiply than to divide as required by the tangent function.

The run of the jack rafter in inches is converted to units by dividing by 12″. In this example $33.34'' \div 12'' = 2'\text{–}9\frac{5}{16}''$, or 2.78 units of run.

The line length of the rafter is then calculated by multiplying the run by the unit length of a common rafter. In this roof with a 6″ unit rise, the jack rafter length is 13.42″ × 2.78, or 37.31″.

Because the run of the jack rafter was established from the plate to the edge of the hip rafter along the long point side of the jack rafter, the calculated length is the actual length and can be marked on the rafter stock as shown in Fig. 6-11 (a). The seat cut and tail of the jack rafter are identical to that of the common rafter.

Figure 6-11 (a)
Jack rafter layout.

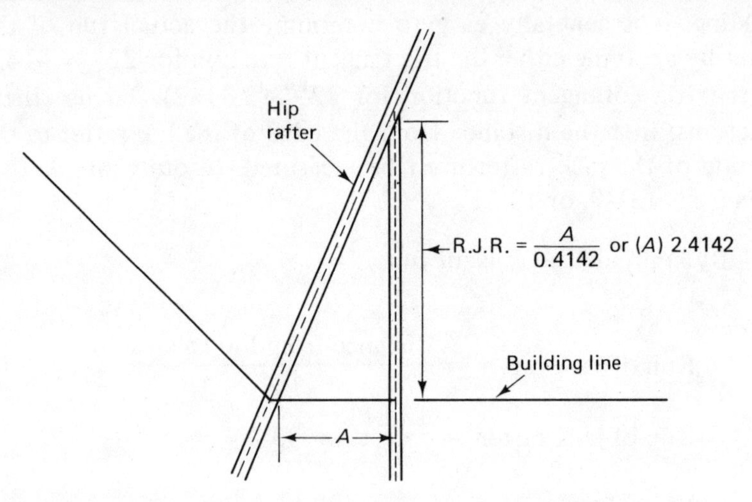

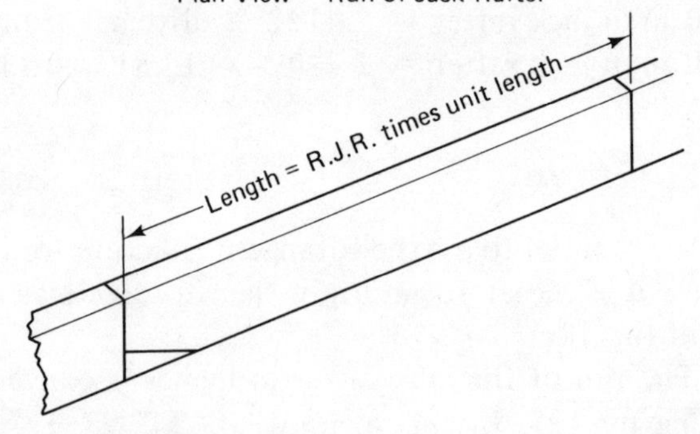

Jack Rafter Bevel. The depth of the bevel for the jack rafter is measured in plan view from the long point to a line squared across the top of the rafter at the short point [see Fig. 6-11 (a)]. For rafter stock $1\frac{1}{2}''$ thick, the depth of the bevel is $3\frac{5}{8}''$. This distance is measured at right angles to the plumb line on the jack rafter and a new plumb line is established. By squaring a line across the top of the rafter, the bevel may be established on the top edge of the rafter stock [see Fig. 6-11 (b)].

Figure 6-11 (b)
Jack rafter layout
(contd.)

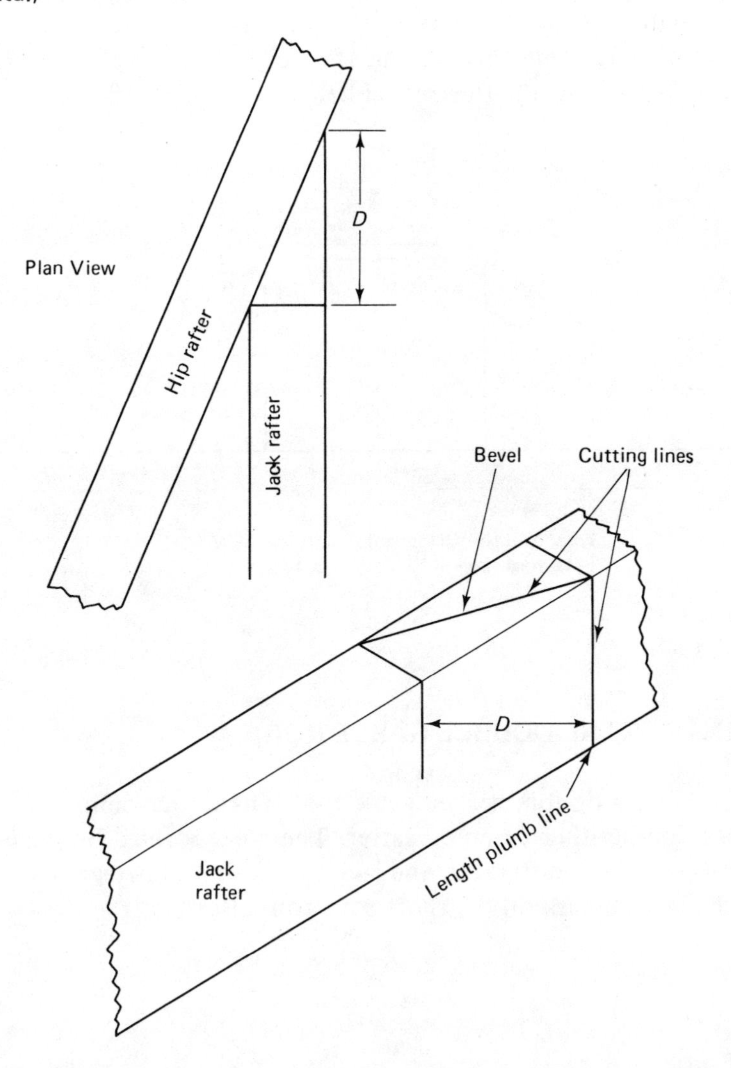

DECK ROOF

The deck roof is a variation of the gable roof in which the rafters do not rise to a peak but instead terminate at a flat deck (see Fig. 6-12). To determine the run of the rafters, the width of deck is subtracted from the overall width of the building. This remaining distance is the actual span of the common rafters. As in framing regular gable roofs, the span is divided by 2 to determine the run.

The building in Fig. 6-12 is 36′ wide with a 10′ central deck. This leaves a rafter span of 26′ and a run of 13′. The theoretical height of the deck above the plate is determined by multiplying the number of units of run by the unit rise in inches. The actual height is determined by adding the height of the seat cut of the common rafter to the theoretical height (see Fig. 6-12).

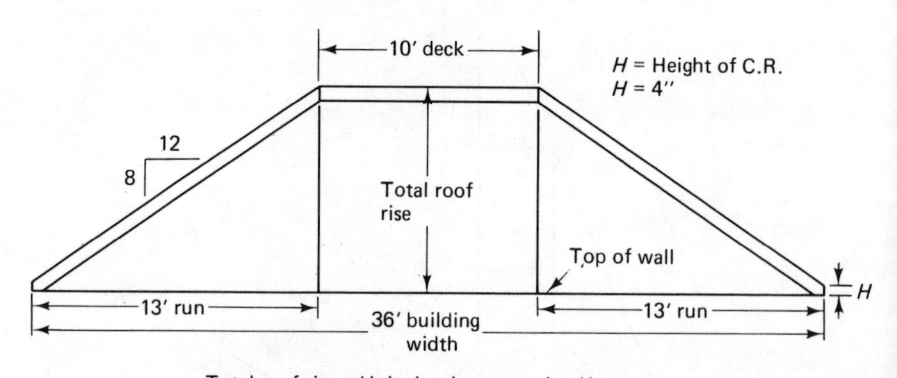

Figure 6-12
Deck roof.

ORNAMENTAL DOUBLE GABLE ROOF

When a double peaked gable roof is used, one common rafter "dies" on another common rafter. The angle formed by the intersection of these rafters may be laid out by first drawing a level line at the length mark on the short common rafter.

Then a plumb line is drawn through the midpoint of the level line, and the points between the length mark and the lower end of the plumb line are connected with a line (see Fig. 6-13). This line is the cutting line and is at the proper angle to allow the short rafter to rest solidly on top of the supporting rafter.

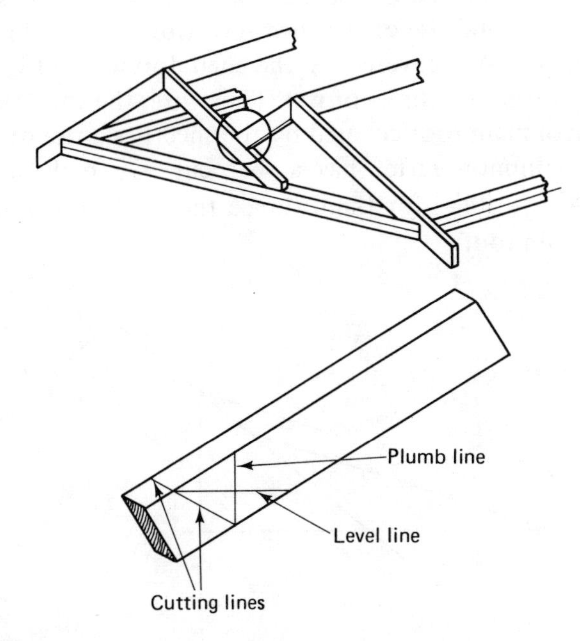

Figure 6-13
Double gable roof.

SHED DORMER

The shed dormer is often used to provide additional room space on the upper floor and is easily framed as explained in Chapter 2 under Shed Dormer. Occasionally, the shed dormer is used for ornamental purposes. When this is done, the upper end of the shed roof rafters rests on the roof sheathing and must be cut at the proper angle.

The length of the rafter is determined in the same manner as for any other common rafter. However, the theoretical line on which the length of the rafter is determined lies on the top edge of the rafter, and the run of the rafter is the horizontal distance from a point directly above the shed wall at the top edge of the rafter to a point directly below the point at the end of the rafter [see Fig. 6-14 (a)].

The distance that the shed roof wall projects above the roof is determined by subtracting the sum of the height of the shed roof seat cut and the total rise of the shed roof from the total rise of the section of main roof covered by the shed dormer. In Fig. 6-14 (a) the shed roof has eight units of run. Therefore, the total rise is 32″. The section of main roof covered by the shed has a rise of 80″. The shed roof common rafter has a seat cut $3\frac{1}{2}''$ high. Therefore, $80'' - 35\frac{1}{2}'' = 44\frac{1}{2}''$ for the distance the shed roof wall projects above the main roof.

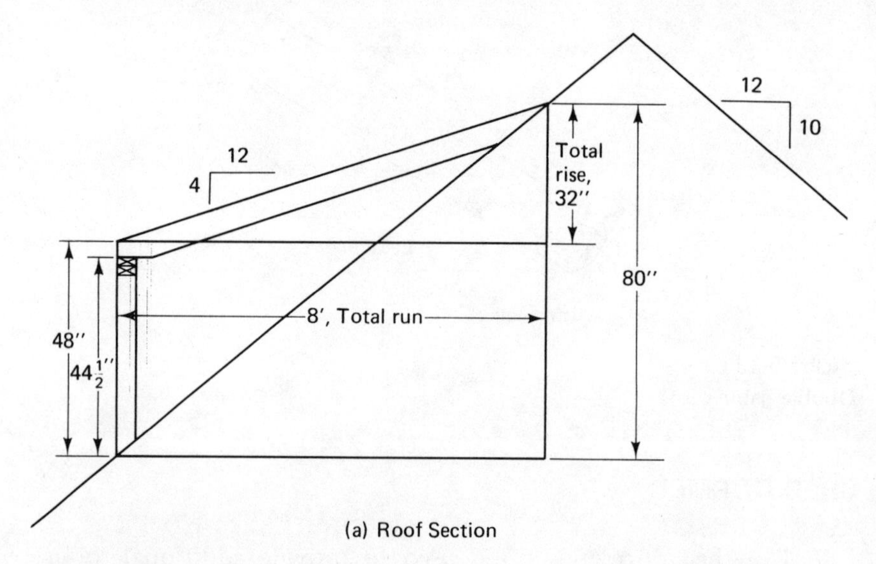

Figure 6-14
Ornamental shed dormer.

To lay out the angle at the upper end of the shed dormer rafter, the framing square is held on the rafter stock with the unit run on the body and the unit rise of the main roof on the tongue [see Fig. 6-14 (b)]. With the square in this position a mark is made at the unit rise of the shed roof and at the unit run. Next, the square is adjusted so that a line can be drawn from the unit run mark through the mark made at the shed roof unit rise [see Fig. 6-14 (c)]. This line represents the cut on the upper end of the rafter.

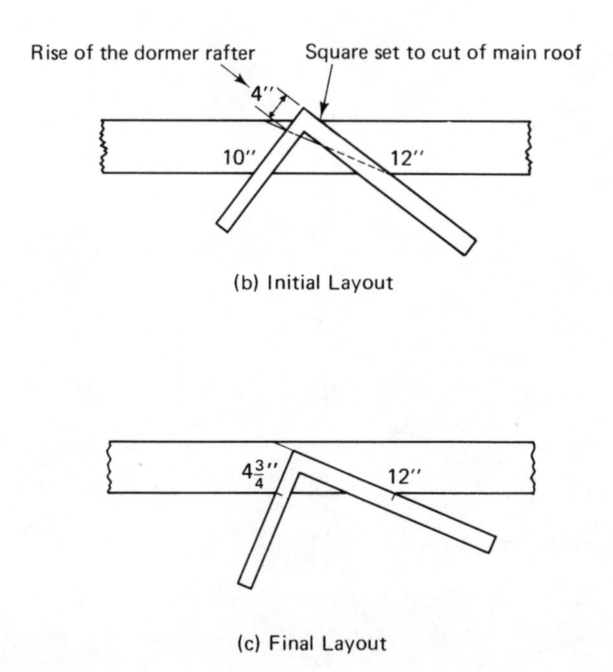

Figure 6-14 (contd.)

GABLE STUDS

Most carpenters prefer to start with the longest gable stud and mark the length of all the remaining studs on this "master pattern." Because studs are placed at a uniform interval such as 16″ or 24″ on center, the difference in length is constant and varies only with the spacing and unit rise. The difference in stud length is determined by multiplying the unit rise by $1\frac{1}{3}$ for studs 16″ O.C.

For studs 24″ on center, the unit rise is multiplied by 2. Table 6-1 gives the difference in length for studs 16″ and 24″ on center for units of rise from 2″ to 18″.

TABLE 6-1

Common Difference in Length of Gable Studs in Inches

Unit Rise	*Spacing: 16″ O.C. Difference in Length*	*Spacing: 24″ O.C. Difference in Length*
2	$2\frac{11}{16}$	4
3	4	6
4	$5\frac{5}{16}$	8
5	$6\frac{11}{16}$	10
6	8	12
7	$9\frac{5}{16}$	14
8	$10\frac{11}{16}$	16
9	12	18
10	$13\frac{5}{16}$	20
11	$14\frac{11}{16}$	22
12	16	24
13	$17\frac{5}{16}$	26
14	18	28
15	20	30
16	$21\frac{5}{16}$	32
17	$22\frac{11}{16}$	34
18	24	36

The angle at the upper end of the gable stud is laid out by holding the unit rise on the tongue of the square and the unit run on the body of the square and marking on the rise side (see Fig. 6-15).

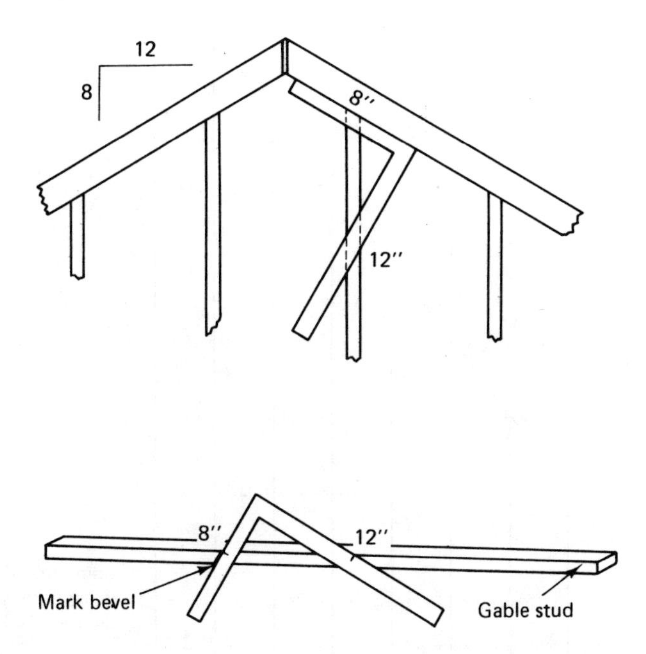

Figure 6-15
Gable studs.

DIFFERENCE IN LENGTH OF JACK RAFTERS FOR ANY ROOF

The common difference in length of jack rafters for any roof can be determined by dividing the number of jack rafter spaces in the common rafter run into the length of the common rafter. This can be illustrated by using the slant triangle section of the roof bordered by the hip rafter, the length of the common rafter, and the common rafter run represented by the plate line (see Fig. 6-16).

The common rafter run represented by the wall plate is divided by the common spacing (center to center distance) of the rafter. In Fig. 6-16 the common rafter run is 12′–8″. Divided by the center distance of 16″, the result is $9\frac{1}{2}$ spaces. Dividing the common rafter length of 15′–10″ by $9\frac{1}{2}$ gives a common difference in length of jack rafters of 20″.

This system can be applied to all rafter spacings and any common rafter length.

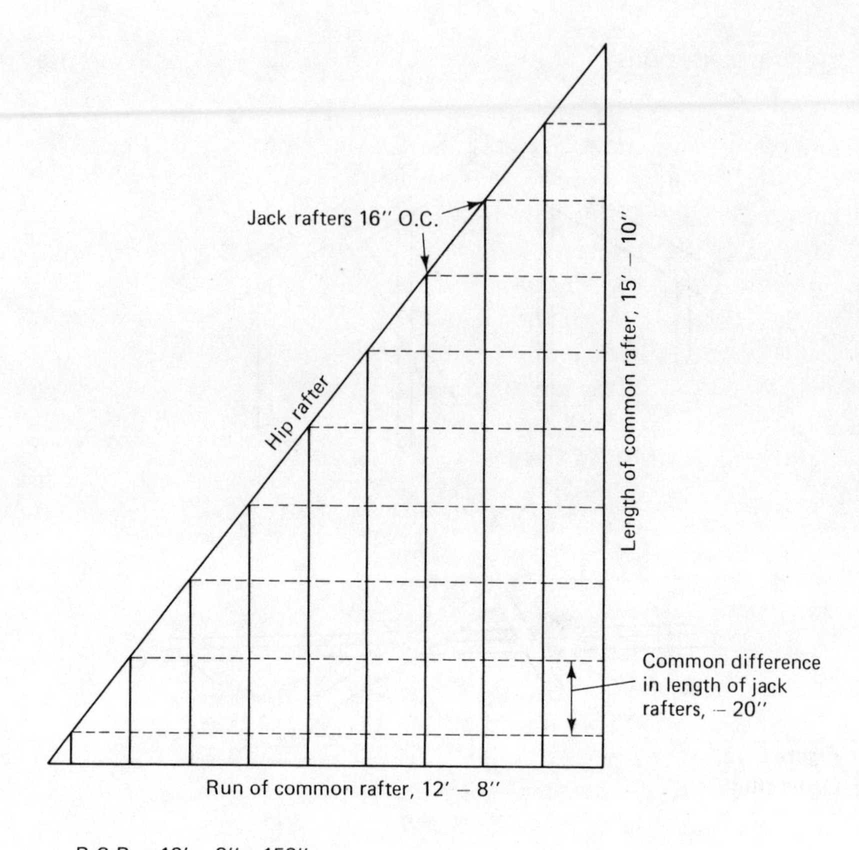

R.C.R. = 12' – 8" = 152"

L.C.R. = 15' – 10" = 190"

$\frac{152''}{16''} = 9\frac{1}{2}$ spaces

$\frac{190''}{9.5} = 20''$ common difference in length of jacks

Figure 6-16
Common difference in length of jacks.

BEVEL LAYOUT FOR ANY JACK RAFTER

The slant triangle application can be used to determine the shape of the bevel for the jack rafters. By making a scale drawing of the slant triangle, a T-bevel may be used to measure the shape of the bevel and to transfer the bevel shape to the jack rafter stock.

The framing square may be placed on the common rafter with the common rafter length represented on the body of the square and the common rafter run represented on the tongue of the square (see Fig. 6-17). It should be noted that in this position the body of the square lies at the same angle as the bevel on the jack rafter. Therefore, the bevels on jack rafters in any pitch roof may be laid out by holding the length of the common rafter on one leg of the square and the run on the other leg of the square and by marking the bevel along the common rafter length side of the square.

When the rafter length is too long to fit on the square (over 24′), the rafter length and rafter run are divided by 2 to obtain figures that will fit on the square.

Figure 6-17
Bevels for jack rafters.

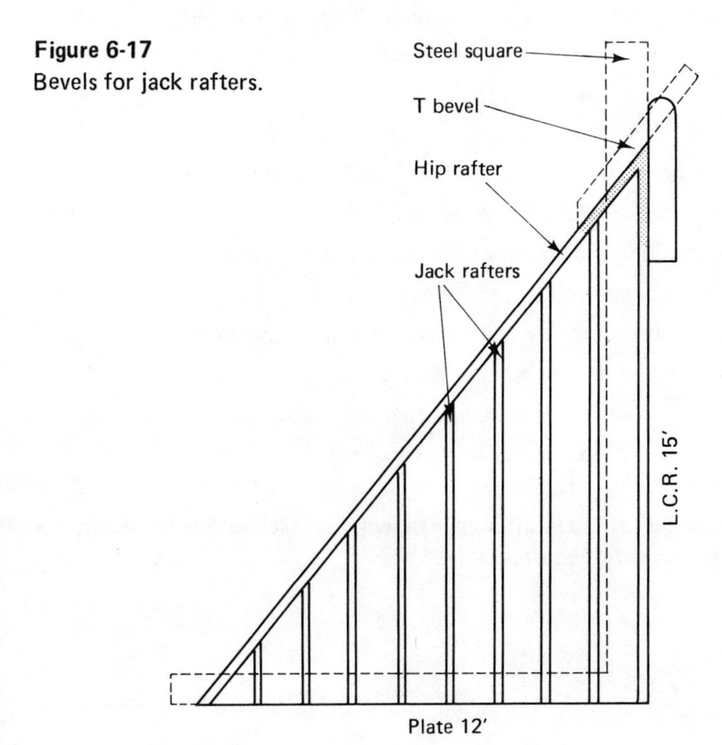

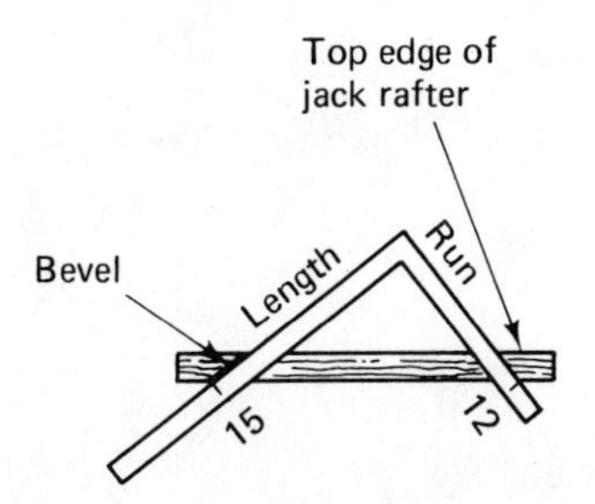

REVIEW QUESTIONS

1. List several unusual roof conditions a carpenter may encounter.
2. What kinds of rafters are used in octagon roofs?
3. What is the angle between the common rafter and the octagon hip rafter?
4. What is the unit of run for octagon hip rafters?
5. What is the tangent of $22\frac{1}{2}°$?
6. Sketch a 4 X 4 ridge layout for an octagon roof.
7. How much shortening is required for octagon hip rafters?
8. How should common rafters be framed in octagon roofs?
9. How is the length of jack rafters calculated in octagon roofs?
10. What is the cotangent of $22\frac{1}{2}°$?
11. How are bevels laid out for octagon roof jack rafters?
12. What is the depth of the jack rafter bevel when the rafter is $1\frac{1}{2}''$ thick?
13. What is a deck roof?
14. How is the run of deck roof rafters determined?
15. How is the height of the deck determined?
16. Sketch the layout for the shortened rafter in an ornamental double gable.
17. Sketch the layout for common rafters in an ornamental shed roof.
18. How is the height of wall for an ornamental shed roof determined?
19. How is the common difference in length of gable studs 16″ O.C. determined? 24″ O.C.?
20. How can the common difference in length of jack rafters be determined without the aid of a rafter table?

APPENDIX

The Framing Square

The steel framing square has become an indispensable carpenter's layout tool and has been manufactured in the United States for over 100 years. Early squares were made from whatever material was available. Today, the carpenter may choose from squares made of aluminum, stainless steel, or plain steel plated or otherwise treated to resist rusting. Aluminum squares are light in weight and will not rust, but the stainless steel square is heavier, stronger, and rustproof. The plated steel square is usually less expensive and is easier to read if it is prevented from rusting.

SQUARE NOMENCLATURE

The steel square, though one piece, consists of two parts: the body and the tongue (see Fig. A-1). The body is 24″ long and 2″ wide. The tongue is 16″ long and $1\frac{1}{2}$″ wide. The heel of the square is the point on the outside edge of the square where the body and tongue meet.

The surfaces of the square are called the *face* and the *back*. The face of the square is the side on which the manufacturer's trademark is stamped. It is the visible side when the body is held in the left hand and the tongue in the right hand. The back of the square is the side opposite the face. It is the visible side when the body is held in the right hand and the tongue in the left hand.

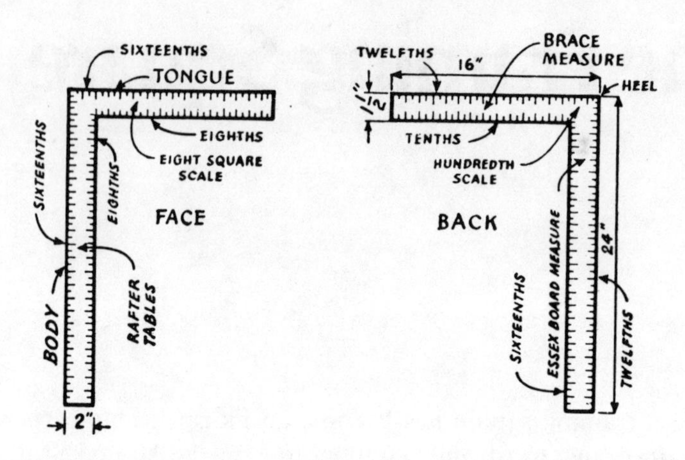

Figure A-1
Square nomenclature.
(Courtesy Stanley Tools)

Scales

The rafter framing square has a number of different measuring scales and tables on the various edges and surfaces. The outside edges of the body and tongue face are divided into inches and sixteenths of an inch. The inside edge is divided into inches and eighths of an inch.

On the back of the square the outside edges of the body and the tongue are divided into inches and twelfths of an inch. The inside edge of the body is divided into inches and sixteenths of an inch, but the inside edge of the tongue is divided into inches and tenths of an inch.

The hundredth scale is located on the back of the tongue at the heel of the square. It is 1″ divided into 100 parts with convenient markings to indicate 25 hundredths and 5 hundredths. The inch divided into sixteenths is located below the hundredths scale on many squares. This makes it easy to convert hundredths to sixteenths "at a glance" (see Fig. A-2).

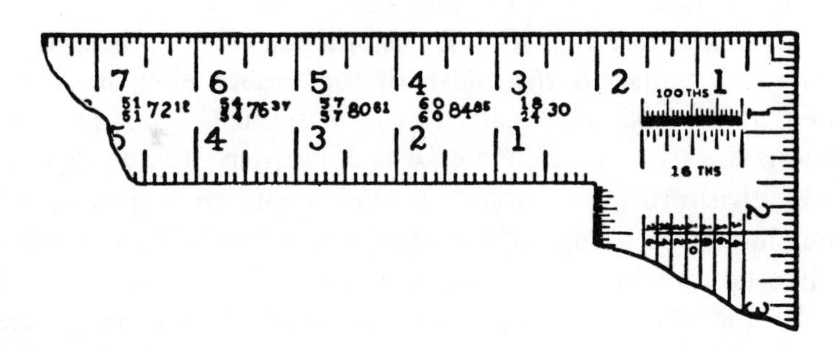

Figure A-2
Hundredth scale.
(Courtesy Stanley Tools)

TABLES

The rafter framing square contains a rafter framing table, an octagon scale, a brace table, and the Essex board measure table. The rafter framing table consists of six lines of information (see Fig. 2-7). The first line of the table gives the length of the common rafter per foot of run. The second line gives the length of hip or valley rafters per foot of common rafter run.

The third line gives the common difference in length for jack rafters spaced 16″ on center. The fourth line gives the common difference in length for jack rafters 24″ on center.

The fifth line gives the side or bevel cuts for jack rafters. To use this line, the number given is used on the body of the square and 12″ is used on the tongue of the square. Always mark the rafter side cut on the tongue of the square. When marking the angle on roof sheathing, always mark on the body side of the square. The sixth line on the rafter framing table gives the side or bevel cut for hip and valley rafters. As with jack rafters, 12″ is used on the tongue of the square, and the bevel is always marked on the tongue.

Octagon Scale

The octagon, or eight scale, is used to shape four-sided material into eight-sided material. The first step in octagon layout when using the octagon scale is to draw a center line across the square in both directions (see Fig. A-3). By using a pair of dividers, a number of divisions equal to the width of the timber in inches can be picked off the octagon scale. If the material is 8″ square, eight divisions are used. If the material is $3\frac{1}{2}''$ square, the divider is set for $3\frac{1}{2}$ divisions. This distance is marked off on each side of the center line at each edge of the square material. By connecting the newly established points, the octagon layout is completed.

In Fig. A-3 the proper distance from the octagon scale is marked off on each side of the center line and reference points *a, b, c, d, e, f, g,* and *h* are established. By connecting points *b* and *c, d* and *e, f* and *g, h* and *a* the octagon is outlined.

Figure A-3
Octagon scale.
(Courtesy Stanley Tools)

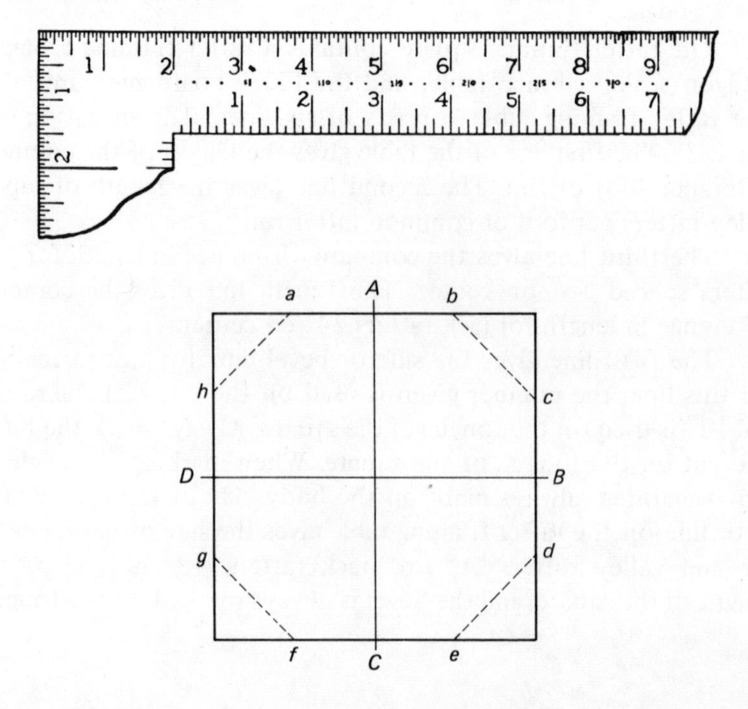

The Brace Table

The brace table on the back of the tongue gives the length of a number of 45° braces (see Fig. A-4). In each group of three

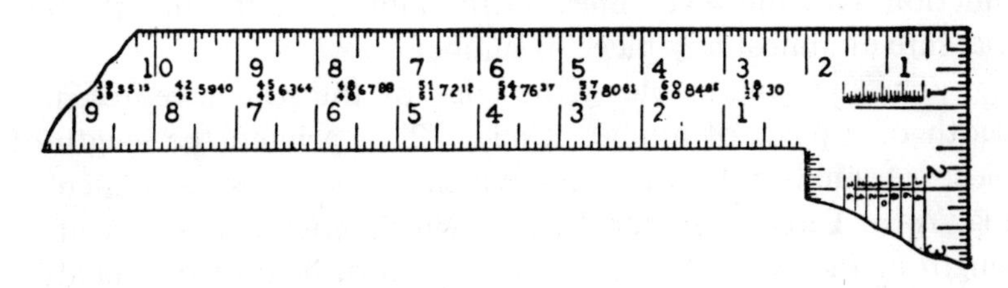

Figure A-4
Brace table. (Courtesy Stanley Tools)

numbers, the numbers on the left represent the legs of the brace triangle. The number on the right represents the length of the brace. In this example taken from the brace table $\frac{39}{39}$ 55.15, the 39″ represent the vertical and horizontal legs of the brace. The 55.15″ are the length of the brace and for practical purposes can be rounded off to $55\frac{1}{8}$″ (see Fig. A-5). The angle on the brace is laid out by holding 12″ on each leg of the framing square and marking along the side outlining the angle in the proper direction.

When the brace has unequal legs, it may be treated as a common rafter. The length of the brace may be determined by applying the figures on the common rafter table, and the top and bottom cuts on the brace are laid out as plumb and level cuts on a common rafter.

Figure A-5
Brace. (Courtesy Stanley Tools)

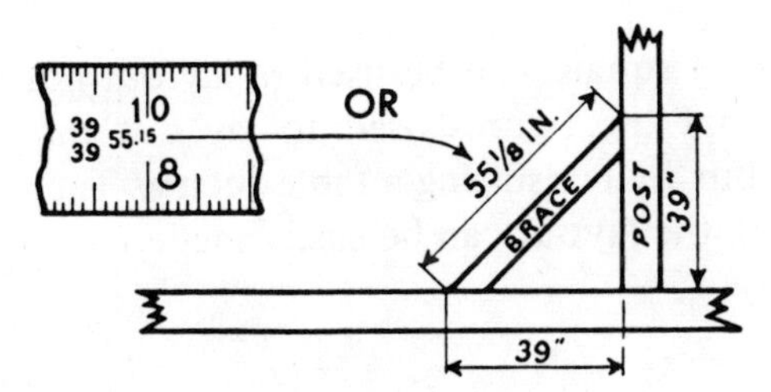

Essex Board Measure

This table is found on the back of the body of the square. The inch marks along the outer edge of the square are used in conjunction with the seven lines on the table to determine the board measure of almost any piece of lumber.

The inch marks along the edge of the square represent the width of a piece of lumber. Under 12″ is printed the length of a piece of lumber 1″ thick for which the scale is calculated (see Fig. A-6). Directly under 12 is 8 which not only represents the length of the board but also the number of board feet included in a board 1″ thick, 12″ wide, and 8′ long.

Under 11″ we read 7|4. This means that there are $7\frac{4}{12}$ board feet of lumber in a board 1″ thick, 11″ wide, and 8′ long. If the board were 2″ thick the 7|4 would be doubled.

To determine the amount of lumber in a 20′-board, the figure for 10′ is doubled. Other lengths not given on the Essex table are handled in the same manner.

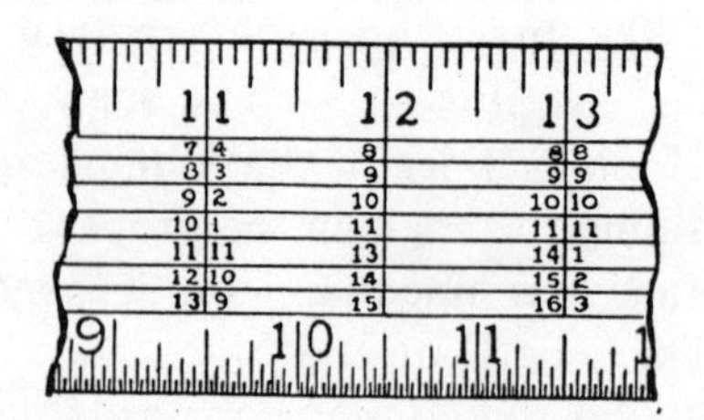

Figure A-6
Essex board measure.
(Courtesy Stanley Tools)

POLYGON MITERS

The framing square can be used to lay out polygon miters. It is not likely that the figures used to lay out the miters can be remembered, but by consulting a table such as Table A-1, Table of Polygon Miters, the layout can be easily made.

TABLE A-1
Table of Polygon Miters

(mark on tongue)

Number of Sides	*Body*	*Tongue*
3	12	$20\frac{3}{4}$
4	12	12
5	12	$8\frac{3}{4}$
6	12	$6\frac{15}{16}$
7	12	$5\frac{3}{4}$
8	12	5
9	12	$4\frac{3}{8}$
10	12	$3\frac{7}{8}$
11	12	$3\frac{1}{2}$
12	12	$3\frac{1}{4}$
13	12	$2\frac{15}{16}$
14	12	$2\frac{3}{4}$
15	12	$2\frac{9}{16}$

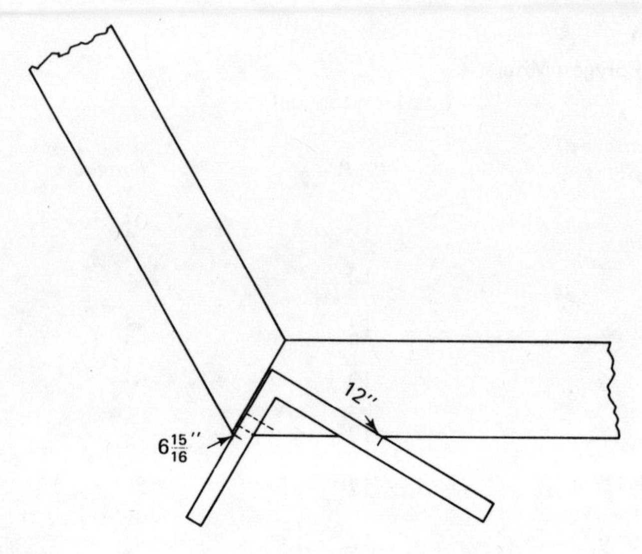

Figure A-7
Polygon layout.

To lay out a six-sided figure (regular hexagon), the length of the sides is marked on the stock and the square is held with 12" on the body, $6\frac{15}{16}$" on the tongue, and a line is marked on the tongue side of the square (see Fig. A-7).

The layout for other polygons is made in a similar manner.

Index

Index

E

F

G

H

I

J

L

M

O

P

R

S

T

U

V